Arthur Cayley

On the Determination of the Orbit of a Planet from Three Observations

Arthur Cayley

On the Determination of the Orbit of a Planet from Three Observations

ISBN/EAN: 9783742868718

Manufactured in Europe, USA, Canada, Australia, Japa

Cover: Foto ©berggeist007 / pixelio.de

Manufactured and distributed by brebook publishing software
(www.brebook.com)

Arthur Cayley

On the Determination of the Orbit of a Planet from Three Observations

ON THE

DETERMINATION OF THE ORBIT OF A PLANET FROM THREE OBSERVATIONS.

BY

ARTHUR CAYLEY, M.A. F.R.S.

SADLERIAN PROFESSOR OF PURE MATHEMATICS,
CAMBRIDGE.

From the Memoirs of the Royal Astronomical Society, Vol. XXXVIII., pp. 17–111.

LONDON:
PRINTED BY STRANGEWAYS AND WALDEN,
CASTLE STREET, LEICESTER SQUARE.
1870.

II. *On the Determination of the Orbit of a Planet from Three Observations.* By Professor Cayley.

Read December 10, 1869.

I propose to consider from a geometrical point of view the problem of the determination of the orbit of a planet from three observations. The orbit is a conic, having the Sun for a focus; and each observation shows that the planet is at the date thereof in a given line. We have thus a given point or focus S, and three given lines, say the "rays." The orbit-plane, if known, would, by its intersections with the three rays, determine the three positions of the planet; that is, we should have the focus and three points on the orbit; or (what is the same thing) three radius vectors from the focus, say a "trivector." Geometrically, through three given points, and with a given focus, there may be described four conics; but (as will be explained) there is only one of these which can be the orbit; we may therefore say that the orbit will be determined, and that uniquely, by means of a given trivector. The problem is therefore to find the orbit-plane, such that in the orbit determined by means of the trivector the times of passage between the three positions on the orbit may have the observed values; or (what is the same thing) that the orbital areas, each divided by the square root of the latus rectum, may have given values. If, instead of the orbit-plane, we consider the orbit-axis (that is, the line normal to the orbit-plane at the point S), or, what is more convenient, the orbit-pole, or intersection of the axis with a sphere about the centre S; then to a given position of the orbit-pole, there corresponds, as above, a determinate orbit; and the problem is to find the position of the orbit-pole, so that in the orbit belonging thereto the times of passage may have given values as already mentioned; and it is clear that the required position of the orbit-pole may be obtained as the intersection of two spherical curves; the one of them, the locus of those positions of the orbit-pole for which the time of passage between the first and second points on the orbit has its proper given value; the other of them, the locus of those positions for which the time of passage between the second and third points on the orbit has its proper given value: and in connexion therewith we may consider other isoparametric loci of the orbit-pole; for instance, the isoeccentric lines, or loci of the orbit-pole such that along each of them the excentricity of the orbit has a given value. It is in this point of view that the problem is considered in the present memoir, viz., the object proposed is the discussion of the configuration, &c. of these loci. I consider, in the first instance, any three given rays whatever; but in the ulterior discussion of the spherical curves, which it is difficult to carry out otherwise than numerically, I have confined myself to the case of a particular symmetrical position of the three rays; viz., these are taken to be lines each of them at an inclination of 60° to a fixed plane through S, and such that their projections on this plane form an equilateral triangle having S for its centre, and that each ray cuts the plane in the mid-point of the corresponding side of the triangle.

The general theory as above explained is further developed in the memoir; and I consider the formulæ for the determination of the orbit, &c. by means of a given trivector; those relating to the determination of the trivector obtained as above by means of a variable plane passing through a given point and intersecting three given rays; and lastly, the application to the particular system of three rays already referred to. The Plates refer to this particular system; they are as follow:—

Plate 1. General Planogram for a single ray,	See Nos. 8–10 for explanation of the terms Planogram and Spherogram.
„ 2. Planogram for Meridian 90°–270°,	
„ 3. Planogram for Meridian 0°–180°,	
„ 4. Spherogram for the Excentricity,	
„ 5. Spherogram for the Time.	

Article Nos. 1 to 14. Considerations on the General Theory.

1. As explained in the introduction, we have a point or focus S, and three lines called the "rays." The orbit-plane is any plane through S; it

meets the rays in three points, which are points on the orbit; and joining these with S, we have a "trivector." The orbit is for the present considered as in general uniquely determined by means of the trivector.

2. There are certain critical positions of the orbit-plane.

First, the orbit-plane may be parallel to one of the rays; or (what is the same thing) it may pass through the line through S parallel to the ray: the point on the ray is at infinity; or say that it is at an indefinitely great distance in one direction or in the other direction along the ray; and (from the particular way in which the orbit is selected as one of four conics) there is, as will appear (see *post*, No. 20), a discontinuity of orbit as the point passes from the one to the other of these positions.

3. *Secondly*, the orbit-plane may be parallel to two of the rays; or (what is the same thing) it may pass through the lines through S parallel to these two rays; the points on the two rays are each at infinity; viz. each of them is at an indefinitely great distance in one or the other direction along the ray; and there is a discontinuity of the orbit as each point passes from the one to the other of its two positions.

4. *Thirdly*, the orbit-plane may be such that the orbit is a right line. To see how this arises, observe that we may consider a system of lines meeting each of the three rays, and of course generating a hyperboloid; say these are the generating lines: there is on the hyperboloid another system of lines, say the directrix lines, in which are included the three rays; the point S is not on the hyperboloid. Then, if the orbit-plane pass through a generating line, it will meet the three rays in the points in which these are met by the generating line: and the orbit is, consequently, the generating line (described, as being a right line not passing through S, with a velocity $= \infty$). Any plane through S and a generating line also meets the hyperboloid in a directrix line; and consequently touches it at the intersection of the two lines, viz. it is a tangent plane of the hyperboloid. The planes in question thus envelope the circumscribed cone whose vertex is S; or (what is the same thing) when the orbit-plane is any tangent-plane of this cone, the orbit is a right line.

5. The only exception is, *fourthly*, when the orbit-plane passes through one of the rays. Observe that the plane then meets the hyperboloid in another line, that is, a generating line, or the case under consideration is included in the third case; it is also included in the first case. The point

on the ray in question is here not a determinate point, but any point whatever of the ray; the points on the other two rays being (as in general) determinate: the orbit is consequently indeterminate; viz. to any point selected at pleasure as the intersection of the orbit-plane with the ray contained therein, there corresponds a determinate orbit (in particular, the selected point may be such that the orbit is, as in the third case, a right line); and, corresponding to the position in question of the orbit-plane, we have the entire system of such orbits.

6. Consider now the corresponding positions of the orbit-pole on a sphere described about the centre S. It will be convenient for the moment to attend to the two opposite positions of the orbit-pole belonging to any position of the orbit-plane, and thus to regard the orbit-pole as moving over the entire spherical surface. The parallel through S to a ray meets the sphere in two points, poles of a great circle which I call a "separator;" we have thus three separators, each two meeting in a pair of opposite points which I call the points B; viz., these are the intersections with the sphere of a line through S perpendicular to the plane containing the parallels of the two rays. A line through S perpendicular to the plane through a ray meets the sphere in a pair of opposite points which I call the points A; these lying on the corresponding separator; there are thus three pairs of points A. The cone reciprocal to the circumscribed cone (that is, generated by a line through S at right angles to any tangent plane of the circumscribed cone) meets the sphere in a spherical conic which I call the "regulator;" this touches each of the separators at the pair of points A on such separator.

7. I say that in the *first* of the cases above considered the locus of the orbit-pole is a separator; in the *second* case the orbit-pole is a point B; in the *third* case the locus is the regulator; and in the *fourth* case the orbit-pole is a point A.

8. In the absence of models, the spherical figure must be represented by a projection; the stereographic projection is convenient for facility of description; and it has the very great advantage that we can by means of it exhibit, no matter how large a portion of the spherical surface. In the figures called "spherograms," afterwards referred to, the representation of a hemisphere is all that is required; but, to give a more distinct general idea, I annex a figure representing a larger portion of the surface; the data

are those belonging to the particular symmetrical case referred to as intended to be specially considered: and the regulator conic is accordingly a pair of opposite small circles, the points A and B being related to it symmetrically; but, disregarding these specialities, the figure is adapted to

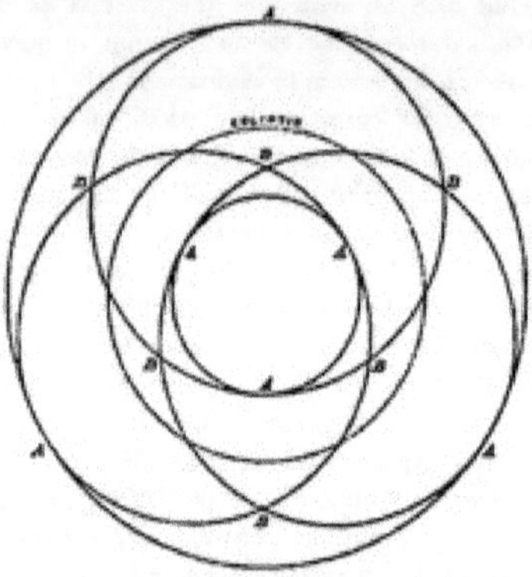

Fig. 1.

the illustration of the general case (at least if the point S be situate *within* the hyperboloid), and it is here given for that purpose. The circle marked "Ecliptic" does not properly belong to the figure: it is added as showing the boundary of a hemisphere, so that, by omitting all that lies outside this circle, the figure would be limited to the representation of a hemisphere; and the orbit-pole be in every case represented, no longer as a pair of opposite points, but as a single point; we should have the separators each as a half circle, and the regulator as a single small circle; the separators would intersect in pairs, in the *three* points B, and would touch the regulator in the *three* points A, &c.

9. The figure constructed as above, but omitting so much of it as lies outside the ecliptic circle, is the representation of a hemisphere—say of the northern hemisphere. It is readily seen that the central triangle BBB and the three circumjacent triangles BBB, represent also the half-surface

of the sphere, viz., instead of the omitted portions of the northern hemisphere we have the equal opposite portions of the southern hemisphere. The adoption of this figure as the representation of the half-surface of the sphere has the great advantage that the spherical curves can be delineated without the apparent breaks which would otherwise occur at their intersections with the ecliptic circle: I accordingly adopt it, and call the figure in question (viz., that composed of the four triangles) a blank "spherogram." We wish for any given position thereon of the orbit-pole to determine the values of certain parameters (excentricity, latus rectum, time of passage between two rays, &c., as the case may be) belonging to the orbit, with a view to the subsequent delineation of the corresponding isoparametric (iseccentric, isochronic, &c.) lines, so constructing a "spherogram" for any such parameter, or system of lines.

10. It is for this purpose convenient to consider the values of the parameter corresponding to a single series of positions of the orbit-pole, viz., we consider the orbit-pole as describing on the sphere a curve selected at pleasure. Consider for a moment the orbit-plane as a material plane rigidly connected with the orbit-axis; the motion of the orbit-pole does not absolutely determine the motion of the orbit-plane, inasmuch as the orbit-plane, occupying the same position in space, might rotate about the orbit-axis; but if we exclude any such motion by the assumption that the motion of the orbit-plane is always about an axis in the orbit-plane, then the motion of the orbit-pole determines that of the orbit-plane, viz., the orbit-plane envelopes a cone, the reciprocal to that described by the orbit-axis. If then on the orbit-plane in each position thereof we mark, as well its line of contact with the enveloped cone, as also its intersections with the three rays, we obtain a figure (which may, if we please, be regarded as drawn on the orbit-plane in some particular position thereof), such figure consisting of a series of trivectors, and (belonging to each of them) a line through S serving to fix the position of the trivector in space. The locus of each extremity of the trivector is a certain curve, and the construction establishes a point-to-point correspondence between these three curves; viz., to any point on one of them there corresponds on each of the other two a single point, the three points being the extremities of a trivector. The figure would be rendered more complete by drawing the orbit be-

longing to each trivector thereof. Such a figure (with or without the orbits) is termed a "planogram."

11. The most simple case is when the orbit-pole describes a great circle; the orbit-plane here rotates about a fixed line, the axis of the circle, or (what is the same thing) the enveloped cone reduces itself to this axis of rotation; and the line of contact is thus a fixed line in the orbit-plane; or (what is the same thing) the lines through S in the planogram are here a single fixed line, the axis of rotation. I say that, for each extremity of the trivector, the locus is a hyperbola, having the axis of rotation for its conjugate axis. In fact, attending to any one ray, it is the same thing whether the orbit-plane be made to revolve round the axis of rotation, so as continually to intersect the ray, or whether, considering the orbit-plane as fixed, and the ray as rigidly connected with the axis, we make the ray to rotate about this axis, so as continually to intersect the orbit-plane. But in this last case the ray describes about the axis a hyperboloid of revolution, and the orbit-plane, as an axial plane, meets this surface in a hyperbola having the axis for its conjugate axis; which hyperbola is the required locus of the trivector-extremity. It is moreover easy to see that if the angle of position of the variable orbit-plane, or (what is the same thing) the angle of position of the orbit-pole in the great circle which it describes be $= q$ (where q is measured from any fixed plane or point), and if the co-ordinates x' and y' be measured from S in the direction of and perpendicular to the axis of rotation, then the co-ordinates of the point on the hyperbola are expressed in the form $x' = \alpha + a \tan (q + \beta)$, $y' = b \sec (q + \beta)$, where α, a, b, β, are constants depending on the position of the ray in regard to the axis of rotation: see as to this *post*, No. 49.

12. Considering the orbit-pole as describing a given curve, the value for the several positions thereof of any parameter of the orbit may be exhibited by means of a "diagram," viz., we may take for abscissa any quantity serving to fix the position of the orbit-pole on the described curve, and for ordinate the value of the parameter in question. In the particular case where the orbit-pole describes a great circle passing through the axis of the stereographic projection, and which is consequently in the spherogram represented by a diameter of the ecliptic or bounding circle, it is natural to take for the abscissa the distance (from the centre) of the representation of the orbit-pole; the diagram will then fit on to the

diameter, and for any position of the orbit-pole on such diameter give at once the value of the parameter to which the diagram relates.

13. It is right to remark that the construction of planograms and diagrams is merely subsidiary to that of the spherograms; the information given by any number of planograms or diagrams would be all of it embodied in a spherogram for the same parameter. And theoretically the construction of a spherogram is a mere matter of geometry; for a given position of the orbit-pole we construct the trivector, thence the orbit, and in relation thereto any parameters which it is desired to consider; and so, for a sufficient number of points on the spherogram, determine the value of the parameter, or parameters; and lay down the isoparametric lines. The construction of the orbit from a given trivector, and in particular the selection of the orbit as one of the four conics given by the trivector, has not yet been explained: in connexion herewith we have the discontinuity of orbit which arises when the orbit-pole is upon a separator, and which is a leading circumstance in the theory; until it is gone into, there is little more to be said in the way of general explanation as to the spherogram, or the isoparametric lines thereof.

14. It may however be noticed that for any parameter whatever, the points A of the spherogram are common points, through which pass in general the lines belonging to any value whatever of the parameter; the reason of course is that the orbit-plane then passing through the ray, and the orbit itself being indeterminate, the value of any parameter belonging to the orbit is also indeterminate. Moreover, for some parameters the curve belonging to any particular value of the parameter not only passes through the points A, but passes through each point twice, or (what is the same thing) has each of the points A for a nodal point; when this is so, then it is to be further observed that, for certain values of the parameters, they will be acnodal points, properly belonging to the curve, although there is not any real branch of the curve passing through the points A; for others they will be crunodal points, with two real branches through each; and in the transition between the two cases they will be cuspidal points on the isoparametric curve; it will appear in the sequel that this is really the case in regard to the isecentric lines.

Article Nos. 15 to 30. Determination of the Orbit from a given Trivector.

15. With a given point S as focus, and through three given points, that is with a given trivector, there may be described *four* conics. This appears from the general theory according to which a given focus is equivalent to two given tangents; and also from the geometrical construction, *Principia*, book i. sect. 4; Scholium to Prop. xxi.: viz. given the focus S and the points 1, 2, 3, then if

On 23 we find a so that $a2 : a3 = S2 : S3$,
" 31 " b " $b3 : b1 = S3 : S1$,
" 12 " c " $c1 : c2 = S1 : S2$,

the points a, b, c, are each of them on the directrix, so that any two of them determine the directrix. In the figure (as in NEWTON's) the distances S 1, S 2, S 3, are each regarded as positive, but the very same construction, taking two of the distances each as positive and negative successively, would lead to three other positions of the directrix; or the construction would give in all four conics.

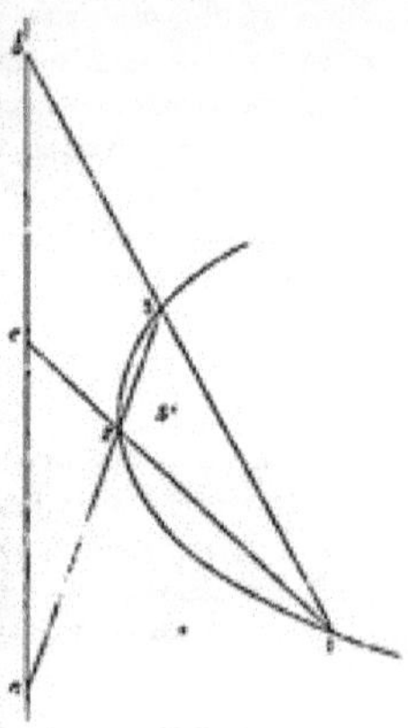

Fig. 1.

16. In the figure the directrix lies on the same side of the three points; and the conic is thus an ellipse or parabola, or, if a hyperbola, then the three points lie in the same branch thereof; and it is consequently an orbit such that along it a body can pass through the three points successively. The construction as varied would give in each case a directrix having on one side of it one, and on the other side two, of the three points; so that the conic would be a hyperbola having the three points not on the same branch thereof; consequently it would not be an orbit such that along it a body could pass through the three points successively.

And it thus appears that though the trivector really determines four conics, yet it is only one of these in which the directrix lies on the same side of the three points; and this conic I call the "orbit:" the given trivector thus determines a single orbit.

17. It is to be noticed however that the orbit constructed as above *may* be a hyperbolic branch separated by the directrix from the focus S, and consequently convex to the focus S; viz., the three points lie here in a hyperbolic branch convex to S, and which is therefore not an orbit which can be described under the action of an *attractive* force at S: say we have a "convex orbit." I regard this as a real orbit, but the times of passage therein as imaginary, or rather as non-existent, and the case is thus excluded from consideration in the formulæ and figures which relate to the times of passage.

18. The same results are established analytically in a very similar manner, viz., taking the focus for origin and starting from the focal equation

$$r = \mathrm{A}x + \mathrm{B}y + \mathrm{C};$$

then if we take (x_1, y_1), (x_2, y_2), (x_3, y_3), as the co-ordinates of the three given points, and write

$$r_1 = \sqrt{x_1^2 + y_1^2},\quad r_2 = \sqrt{x_2^2 + y_2^2},\quad r_3 = \sqrt{x_3^2 + y_3^2},$$

we have for the determination of the constants

$$\begin{aligned} r_1 &= \mathrm{A}x_1 + \mathrm{B}y_1 + \mathrm{C}, \\ r_2 &= \mathrm{A}x_2 + \mathrm{B}y_2 + \mathrm{C}, \\ r_3 &= \mathrm{A}x_3 + \mathrm{B}y_3 + \mathrm{C}, \end{aligned}$$

and the equation therefore is

$$\begin{vmatrix} r, & x, & y, & 1 \\ r_1, & x_1, & y_1, & 1 \\ r_2, & x_2, & y_2, & 1 \\ r_3, & x_3, & y_3, & 1 \end{vmatrix} = 0,$$

which, attributing therein to r_1, r_2, r_3 the signs $+$, $-$ at pleasure, represents eight different equations: these however give only four conics, viz., we have the same conic whether we attribute to r_1, r_2, r_3 any particular combination of signs, or reverse all the signs simultaneously.

19. But the focal equation $r = \mathrm{A}x + \mathrm{B}y + \mathrm{C}$ is precisely equivalent to the equation

$$r = \frac{p}{1 + e\cos(\theta - \omega)},$$

and in this equation (taking as is allowable p as positive) then if $\pm e$ be $=$ or <1, that is for an ellipse or parabola whatever be the value of $\theta-\varpi$, r is always positive; but if $\pm e$ be >1, that is for a hyperbola, r is positive for those values of $\theta-\varpi$ which belong to one branch, negative for those which belong to the other branch, of the curve. Hence in the determinant equation, unless $r_{,}, r_{,,}, r_{,,,}$ have the same sign, the curve will be a hyperbola with the points two of them on one branch, the third on the other branch thereof. But in the remaining case, when $r_{,}, r_{,,}, r_{,,,}$ have all the same sign, or say when they are all positive, then the conic is an ellipse or parabola, or else it is a hyperbola with the three points on the same branch thereof; that is, the foregoing determinant equation, regarding therein $r_{,}, r_{,,}, r_{,,,}$ as all of them positive, gives the orbit.

20. When one of the points is at infinity on a given line there is a discontinuity of orbit. To explain this, suppose that the point $(x_{,}, y_{,})$ is situate on the line $y = x \tan\alpha$, at an indefinitely great distance $r_{,}$ in one or the other direction along the line; viz., $r_{,}$ is an indefinitely large positive quantity, and we have in the one case $x_{,}, y_{,} = r_{,}\cos\alpha, r_{,}\sin\alpha$; and in the other case $x_{,}, y_{,} = -r_{,}\cos\alpha, -r_{,}\sin\alpha$: the corresponding equations of the orbit, putting therein ultimately $r_{,} = +\infty$, are

$$\begin{vmatrix} r_{,} & x_{,} & y_{,} & 1 \\ 1, & \cos\alpha, & \sin\alpha, & 0 \\ r_{,,} & x_{,,} & y_{,,} & 1 \\ r_{,,,} & x_{,,,} & y_{,,,} & 1 \end{vmatrix} = 0, \qquad \begin{vmatrix} r_{,} & x_{,} & y_{,} & 1 \\ 1, & -\cos\alpha, & -\sin\alpha, & 0 \\ r_{,,} & x_{,,} & y_{,,} & 1 \\ r_{,,,} & x_{,,,} & y_{,,,} & 1 \end{vmatrix} = 0$$

which equations belong, it is clear, to two distinct conics; or as the point $(x_{,}, y_{,})$ passes from a positive to a negative infinity along the given line, there is an abrupt change of orbit. It is proper to remark that the two orbits are the very same as would be obtained by writing $x_{,}, y_{,} = r_{,}\cos\alpha, r_{,}\sin\alpha$, $r_{,} = +\infty$ and $r_{,} = -\infty$ in the determinant equation: that is, the orbit passes abruptly from one to another of the four conics which belong to the position $(x_{,}, y_{,})$, and we thus understand how the transition from $+\infty$ to $-\infty$, which is geometrically no breach of continuity, occasions in the actual problem a discontinuity.

21. The same thing appears from the geometrical construction; and we derive a further result which will be useful. Suppose first that the point 1 is at infinity in the direction shown by the arrow; then drawing

2 c = 2 S and 3 b = 3 S each in the direction opposite to S 1, we have the points b, c on the directrix, which is thus the line D joining these points. But if 1 is at infinity on the same line in the opposite direction, then instead of c, b we have the points c', b', and the directrix is the line D' joining these points.

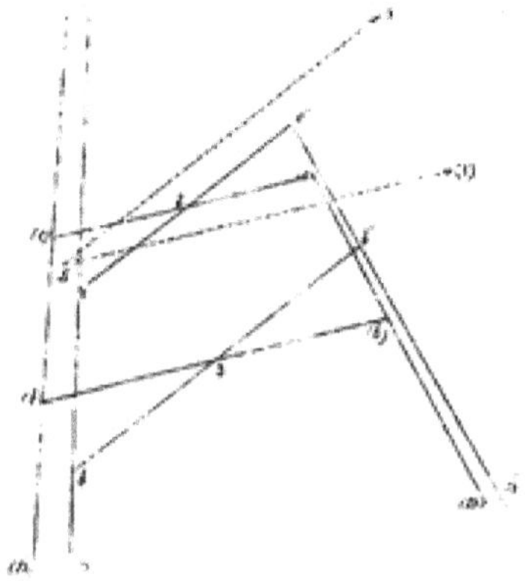

Fig. 3.

22. Observe that in the first case the focus S and the three points are on opposite sides of the directrix D, or the orbit is convex; but in the second case the focus S and the three points are on the same side of the directrix D', and the orbit is concave. That is, the line S_1 does not separate the two points 2, 3, and the orbits are the one convex, the other concave.

23. But if 1 be at infinity along the line S (1) first in the direction shown by the arrow, and then in the opposite direction; in the first case the directrix is (D) not separating the focus S from the three points, and the orbit is concave; in the second case the orbit is (D'), not separating S from the three points, and the orbit is still concave; here the line S (1) does separate the points 2, 3, and the orbits are both concave.

24. And we thus see in general that as the point 1 passes from a positive to a negative infinity along a line passing through S; then,

according as the line through S does not or does separate the remaining two points 2, 3, the orbits corresponding to the two positions of 1 are the one convex, the other concave, or they are both concave.

25. The points 1 and 2 may be each of them at infinity along a given ray; we have here in a similar manner $x_1, y_1 = r_1 \cos \alpha_1, r_1 \sin \alpha_1$, or else $= -r_1 \cos \alpha_1, -r_1 \sin \alpha_1$, where r_1 is an indefinitely large positive quantity; and $x_2, y_2 = r_2 \cos \alpha_2, r_2 \sin \alpha_2$, or else $= -r_2 \cos \alpha_2, -r_2 \sin \alpha_2$, where r_2 is an indefinitely large positive quantity. And writing ultimately $r_1 = +\infty, r_2 = +\infty$, the equation of the orbit is obtained in the form

$$\begin{vmatrix} r, & x, & y, & 1 \\ 1, & \pm \cos \alpha_1, & \pm \sin \alpha_1, & 0 \\ 1, & \pm \cos \alpha_2, & \pm \sin \alpha_2, & 0 \\ r_3, & x_3, & y_3, & 1 \end{vmatrix} = 0,$$

where the $\pm$ of the second line and the $\pm$ of the third line have each of them the value + or − at pleasure. There are consequently four distinct orbits, corresponding to the combinations of each of the two directions of the point 1 with each of the two directions of the point 2. And it is moreover clear that these are the very conics which are obtained from the determinant equation by writing therein $x_1, y_1 = r_1 \cos \alpha_1, r_1 \sin \alpha_1$; $x_2, y_2 = r_2 \cos \alpha_2, r_2 \sin \alpha_2$ and $r_1 = +\infty, -\infty$; $r_2 = +\infty, -\infty$ successively; viz., the orbit changes abruptly between the four conics which correspond to the given position of the points 1, 2, 3.

26. The geometrical construction is very simple indeed; viz., measuring off from 3 in the directions S1, S2, and in the opposite directions respectively, a distance = S3, we have four points, the angles of a rectangle; and joining these in pairs, we have the four positions of the directrix: the figure shows at once that the orbits are three of them concave, the remaining one convex.

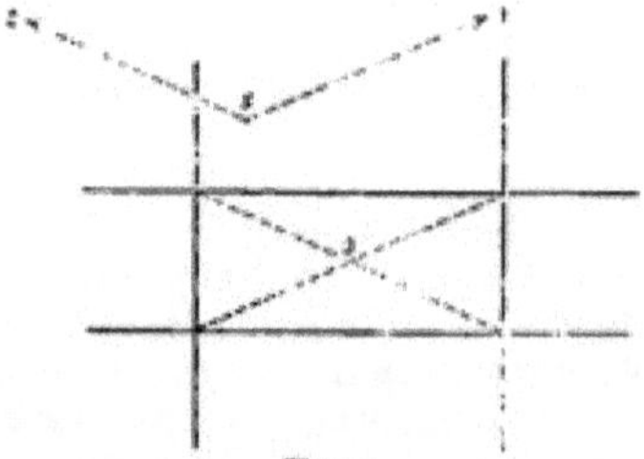

Fig. 4.

27. The determinant equation obtained for the orbit is an equation of the form

$$r = \mathrm{A}x + \mathrm{B}y + \mathrm{C};$$

and it is clear that the equation of the directrix is $\mathrm{A}x + \mathrm{B}y + \mathrm{C} = 0$. By what precedes, this line will lie on the same side of the three points, viz., either it does not separate them from the focus, and the orbit is then concave, or it does separate them from the focus, and the orbit is then convex. Although in general the sign of C is no criterion (for the equations $r = \mathrm{A}x + \mathrm{B}y + \mathrm{C}$ and $r = -\mathrm{A}x - \mathrm{B}y - \mathrm{C}$ represent the same curve) yet in the present case it is so; for, observe that, in taking r_1, r_2, r_3 each of them positive, we make r to be positive for the orbit, that is, for the entire curve if an ellipse or parabola, but for the branch containing the three points if the curve is a hyperbola. Hence, considering the radius vector through S parallel to the directrix, this is positive for a concave, negative for a convex orbit; or writing $\mathrm{A}x + \mathrm{B}y = 0$, we have $r = \mathrm{C}$ positive for a concave, negative for a convex orbit; wherefore the orbit is concave or convex according as C is positive or negative.

28. Comparing the equation with

$$r = e(x\cos\varpi + y\sin\varpi) \pm a(1 - e^2),$$

we see that the excentricity and semiaxis major, taken to be each of them positive, are

$$e = \sqrt{\mathrm{A}^2 + \mathrm{B}^2}, \qquad a = \frac{\pm\mathrm{C}}{1 - \mathrm{A}^2 - \mathrm{B}^2}$$

($+\mathrm{C}$ or $-\mathrm{C}$, according as $e < 1$ or $e > 1$); and inasmuch as the focus and directrix are known, there is no ambiguity as to the position of the orbit: it may be added that the co-ordinates of the centre are given by

$$\begin{aligned}(\mathrm{A}^2 - 1)\,x + \mathrm{A}\mathrm{B}\,y + \mathrm{A}\mathrm{C} &= 0,\\ \mathrm{A}\mathrm{B}\,x + (\mathrm{B}^2 - 1)\,y + \mathrm{B}\mathrm{C} &= 0,\end{aligned}$$

that is, we have for the co-ordinates of the centre

$$x = \frac{\mathrm{A}\mathrm{C}}{1 - \mathrm{A}^2 - \mathrm{B}^2}, \qquad y = \frac{\mathrm{B}\mathrm{C}}{1 - \mathrm{A}^2 - \mathrm{B}^2};$$

and thence also

$$x = \frac{2\,AC}{1 - A^2 - B^2}, \qquad y = \frac{2\,BC}{1 - A^2 - B^2}$$

for the co-ordinates of the other focus.

29. But to effect the comparison rather more precisely it is to be observed that a, e being positive, then for a concave orbit, if X be measured from the focus in the direction *away from* the directrix, we should have

$$r = eX \pm a\,(1 - e^2)$$

(+ for the ellipse, − for the hyperbola, so that $\pm a\,(1 - e^2)$ is positive): whence

$$e = \sqrt{A^2 + B^2}, \quad X = \frac{Ax + By}{\sqrt{A^2 + B^2}}, \quad a = \frac{\pm C}{1 - A^2 - B^2}$$

(by what precedes, C is = +, so that the formula gives as it should do $a = +$).

And similarly for a convex orbit, if X be measured in the direction *towards* the directrix, we should have

$$r = eX - a\,(e^2 - 1);$$

whence

$$e = \sqrt{A^2 + B^2}, \quad X = \frac{Ax + By}{\sqrt{A^2 + B^2}}, \quad a = \frac{-C}{A^2 + B^2 - 1},$$

where by what precedes C is = −, and the formula gives as it should do $a = +$.

30. It is not necessary for the purpose of the present memoir, but I notice an elegant form of the polar equation of the orbit belonging to a given trivector; viz., taking (r, θ) as polar co-ordinates, and therefore (r_1, θ_1), (r_2, θ_2), (r_3, θ_3), as the co-ordinates of the given points, the equation of the orbit is

$$\frac{1}{r} = \Sigma\,\frac{1}{r_1}\cdot\frac{\sin\tfrac{1}{2}(\theta - \theta_2)\sin\tfrac{1}{2}(\theta - \theta_3)}{\sin\tfrac{1}{2}(\theta_1 - \theta_2)\sin\tfrac{1}{2}(\theta_1 - \theta_3)}.$$

In fact, it is clear that this is an equation of the form

$$\frac{1}{r} = (\alpha, \beta, \gamma)(\sin \tfrac{1}{2}\theta, \cos \tfrac{1}{2}\theta)^2:$$

that is of the form

$$\frac{1}{r} = \lambda \cos\theta + \mu \sin\theta + \nu;$$

and that it thus represents a conic with the given focus; and moreover that the equation is satisfied by writing therein (r_1, θ_1), (r_2, θ_2), or (r_3, θ_3), in place of (r, θ); that is, the conic passes through the three given points. The foregoing remarks as to the signs of r_1, r_2, r_3, apply without alteration to this polar equation.

Article Nos. 31 *to* 41. *Time Formulæ;* Lambert's *Equation.*

31. Suppose for a moment that the orbit is an ellipse; as the ellipse may be described in either direction, the time of passage between any two points, 1 to 2, or 2 to 1, indifferently, may be regarded as positive. With only two points 1, 2, we might pass, say from 1 to 2, in either direction along the ellipse, and the time of passage would have ambiguously either of two positive values. In the case however where we have on the ellipse three points, 1, 2, 3, this ambiguity is avoided; viz., it is assumed that the passage between any two of the points is along the elliptic arc which does not contain the third point; the three times of passage are thus all of them positive, and their sum is equal to the periodic time, or time of describing the entire ellipse.

32. But if the orbit be a parabola or concave hyperbolic branch, then, if the points taken in their order of position along the orbit be 1, 2, 3, we have in like manner a positive time of passage between 1 and 2, and also a positive time of passage between 2 and 3; but, inasmuch as there is no passage between 1 and 3 except through 2 (which mode is excluded from consideration), I say that there is no time of passage between 1 and 3; and so consider only two times of passage; viz., between 1 and 2, and between 2 and 3.

33. In the case of a convex hyperbolic branch, since this cannot be described under the action of an attractive force, there is not any time of passage to be considered.

In the transition case of a right line not passing through the focus, since, as mentioned, the velocity is infinite, if the order of the points on the line is 1, 2, 3, the times of passage from 1 to 2 and from 2 to 3 are each $= 0$; and these are the only times of passage which are to be considered.

34. The preceding conventions are of course to be attended to in the application of any formula to the calculation of the times of passage between given points of the orbit; in the case of a parabolic or hyperbolic orbit we have only to ascertain which are the two times of passage to be calculated; but, in the case of an ellipse, we must take care that the time of passage between each two of the three points is calculated along the arc not containing the third point; viz., it is in some cases to be calculated through the angle $< \pi$ between the two radius vectors, and in other cases through the angle $> \pi$ between the two radius vectors; or, more simply, the time to be calculated is sometimes the longer, and at other times the shorter time of passage.

35. For the purpose of the present memoir the unit of time is so fixed that the periodic time in a circle radius 1 shall be equal 2. The periodic in a circle or ellipse, radius or semiaxis major $= a$, is thus $= 2a^{\frac{3}{2}}$, and generally

$$\text{Time} = \frac{2}{\pi} \cdot \frac{\text{Area}}{\sqrt{\frac{1}{2}\text{ latus rectum}}}.$$

The time formulæ are first the ordinary ones in which the time from pericentre is expressed in terms of an angle (the excentric anomaly for an ellipse or hyperbola, true anomaly for the parabola); secondly, LAMBERT's formulæ, in which the time between any two points on the orbit is expressed by means of the two radius vectors and the chord.

36. The first set of formulæ may be written :

Ellipse. u, the excentric anomaly from pericentre, viz. $x = a(\cos u - e)$, $y = a\sqrt{1-e^2}\sin u$, if x, y, are the co-ordinates from the focus, x measured in the direction *towards* the directrix.

$$\text{Time from Pericentre} = \frac{1}{\pi}a^{\frac{3}{2}}(u - e\sin u).$$

Parabola. θ, the true anomaly, viz., $r = p \sec^2 \frac{1}{2}\theta$, if p be the pericentric distance or $\frac{1}{2}$-latus rectum.

$$\text{Time from pericentre} = \frac{3}{\pi}\frac{p^{\frac{3}{2}}}{\sqrt{2}}\left(\tan\frac{1}{2}\theta + \frac{1}{3}\tan^3\frac{1}{2}\theta\right).$$

Hyperbola; concave branch. u, the excentric anomaly from pericentre, viz., $r = a(\sec u - e)$, $y = a\sqrt{e^2 - 1}\tan u$, if x, y are the co-ordinates from the focus, x measured in the direction *away* from the directrix.

$$\text{Time from pericentre} = \frac{3a^{\frac{3}{2}}}{2\pi}\left\{e\tan u - \text{hyp. log}\tan\left(\frac{\pi}{4} + \frac{1}{2}u\right)\right\}$$

and by taking the sum or the difference of two of these expressions, we obtain the time of passage between two given points of the orbit.

37. I remark that as to the elliptic and parabolic orbits, I have preferred using LAMBERT's equations, and I should have done the same for the hyperbolic orbits, but for the absence of a table (see *post*, No. 39). As it is, for the few hyperbolic orbits which it was necessary to calculate, I have used the foregoing formula:* a table of hyp. log tan $\left(\frac{\pi}{4} + \frac{1}{2}u\right)$, $u = 0°$ to $u = 90°$, at intervals of $30'$ to 12 places of decimals, with fifth differences is given, Table IV. LEGENDRE *Traité des Fonctions Elliptiques*, t. ii. pp. 256–259.

38. The other set of formulæ may be written:

Ellipse. r, r', the radius vectors, γ the chord.

$$2a\cos\chi = 2a - r - r' - \gamma, \quad 2a\cos\chi' = 2a - r - r' + \gamma.$$

$$\text{Time} = \frac{3}{2\pi}a^{\frac{3}{2}}(\chi - \chi' - \sin\chi + \sin\chi').$$

Parabola. r, r', γ, *ut supra*;

$$\text{Time} = \frac{1}{4\pi}\left\{(r + r' + \gamma)^{\frac{3}{2}} - (r + r' - \gamma)^{\frac{3}{2}}\right\}.$$

Hyperbola.

$$2a\cosh\chi = 2a + r + r' + \gamma, \quad 2a\cosh\chi' = 2a + r + r' - \gamma.$$

---

* I rather regret that I did not use the foregoing formulæ in all cases.

$$\text{Time} = \frac{1}{2\pi} a^{\frac{3}{2}} (-\chi + \chi' + \sinh \chi - \sinh \chi')$$

where cosh, sinh, denote the hyperbolic cosine and sine of χ, viz.:

$$\cosh \chi = \frac{1}{2} \left(e^{\chi} + e^{-\chi}\right)$$

$$\sinh \chi = \frac{1}{2} \left(e^{\chi} - e^{-\chi}\right).$$

39. The logarithms (ordinary) of the functions $\cosh \chi$, $\sinh \chi$, and of $\tanh \chi$ are tabulated by GUDERMANN, *Crelle*, tt. viii. and ix. from $\chi = 2{\cdot}000$ to $\chi = 8{\cdot}00$ at intervals of ·001 and subsequently of ·01 to eight places of decimals. I do not know why the tabulation was not commenced from $\chi = 0$, but the omission from them of the values 0 to 2 rendered the tables unavailing for the present purpose, and I therefore, for the hyperbolic orbits, resorted to the first set of formulæ.

40. As regards the elliptic formulæ it remains to be explained how the values of χ, χ' are to be selected from those which satisfy the required conditions

$$2a \cos \chi = 2a - r - r' - \gamma, \quad 2a \cos \chi' = 2a - r - r' + \gamma.$$

It is remarked in GAUSS' *Theoria Motûs*, p. 120, that χ is a positive angle between 0° and 360°; χ' a positive or negative angle between + 180°, − 180°, viz. χ' is positive or negative according as the angle between the two radius vectors is < 180° or > 180°. This determines χ', but it is said that χ is really indeterminate; viz. it is so if only the values r, r', γ, a, are given, for there are then two orbits in which these quantities have their given values, and the times in these have different values. But when, as in the case here considered the orbit is known, χ will of course have a determinate signification, and it is easy to explain how this is to be fixed. I observe, in the first place, that if $\chi = \pi$ we have $\gamma = (2a - r) + (2a - r')$, that is, the chord γ passes through the other focus of the ellipse. The criterion thus depends on the position of the two points on the ellipse in relation to the other focus, and it is easy to see that it is as follows: viz. let the time between the points 1, 2, on the ellipse be understood to mean the time of passage from 1 through apocentre to 2; then I say that, in the preceding formula

$$\text{Time} = \frac{1}{2\pi}\, a^{\frac{3}{2}} (\chi - \chi' - \sin\chi + \sin\chi'),$$

χ will be $< 180^\circ$ or $> 180^\circ$ according as the chord from 1 through the other focus H does not or does separate the point 2 from the focus S.

41. It is hardly necessary to remark that in the application of the formulæ, χ, χ' must be reckoned according to their lengths as circular arcs to the radius unity: a table for the conversion of degrees and minutes to such circular measure, is given in most collections of Trigonometrical Tables.

Articles Nos. 42 to 45. Formulæ for the Transformation between two sets of Rectangular Axes.

42. Consider an arbitrary set of fixed rectangular axes, Sx, Sy, Sz, which are considered as intersecting the sphere, centre S, in the points X, Y, Z, and so the axes Sx', Sy', Sz', afterwards defined are considered as intersecting the sphere in the points X′, Y′, Z′. For convenience Sx is considered as an origin of longitudes, which are measured in the plane of xy in the direction towards y; and an angular distance from Sz is termed a polar distance or colatitude; so that the position of any line through S, or point on the sphere, will be determined by its longitude b and colatitude c.

43. It is wished in the sequel to make the orbit-pole revolve about an arbitrary line Sx', and for this purpose I take the new set of rectangular axes, Sx', Sy', Sz', or points on the sphere X′, Y′, Z′, as follows,

$$\text{X}', \text{ longitude G, colatitude } 90^\circ + \text{N}.$$

Y′ Z′, is then a great circle, pole X′, meeting Z X′ in a point Π, longitude G, colatitude N, and the position of Z′ in this great circle is fixed by its distance from Π, ΠZ′ = H, the distance of Y′ being Π Y′ = 90 + H, and these being each of them reckoned from Π in the direction of longitude X to Y. The position of the new axes $S x'$, $S y'$, $S z'$, or points X′, Y′, Z′, is thus fixed by means of the three angles G, N, H.

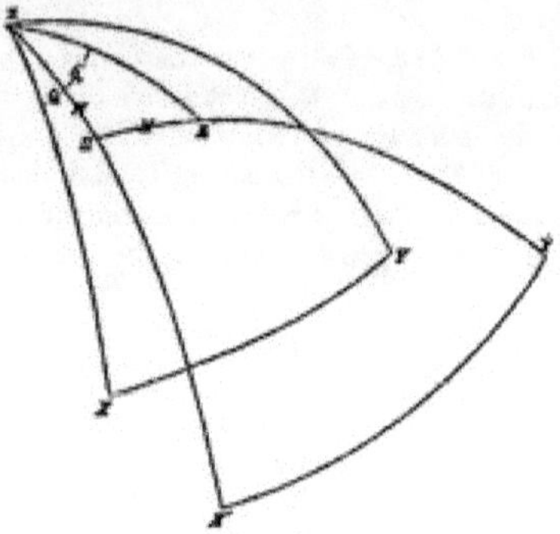

Fig. 5.

It is to be added that if the angle X′ Z Z′ is called φ, and if b, c, are the longitude and colatitude of Z′, then we have $\sin N = \cot \varphi \tan H$, which gives φ, and then

$$b = G + \varphi$$
$$\cos c = \cos N \cos H.$$

44. The transformation-formulæ between the two sets of axes are at once found to be

	X	Y	Z
X′	$\cos G \cos N$	$\sin G \cos N$	$-\sin N$
Y′	$-\sin G \cos H - \cos G \sin H \sin N$	$\cos G \cos H - \sin G \sin H \sin N$	$-\sin H \cos N$
Z′	$-\sin G \sin H + \cos G \cos H \sin N$	$\cos G \sin H + \sin G \cos H \sin N$	$\cos H \cos N$

which are for shortness represented by

	X	Y	Z
X'	α	β	γ
Y'	α'	β'	γ'
Z'	α''	β''	γ''

45. In the particular case where Sx' is in the plane of xy, $N=0$; H coincides with Z, and the longitude and colatitude of Z' are $b = G + 90°$, $c = H$. Writing accordingly in the formula $N = 0$, and introducing b, c in the place of G, H, the formulæ become

	X	Y	Z
X'	$\sin b$	$-\cos b$	0
Y'	$\cos b \cos c$	$\sin b \cos c$	$-\sin c$
Z'	$\cos b \sin c$	$\sin b \sin c$	$\cos c$

and in particular if $c = 0$, (Sz' here coincides with Sz, and the axes Sx', Sy', are in the plane of xy) then we have simply

	X	Y	Z
X'	$\sin b$	$-\cos b$	0
Y'	$\cos b$	$\sin b$	0
Z'	0	0	1

Article Nos. 46 to 60. Application to finding the Intersection of the Orbit-pole by a Single Ray.

46. The equations of the ray referred to the fixed axes are taken to be

$$\frac{x-A}{f} = \frac{y-B}{g} = \frac{z-C}{h}, = R \text{ suppose,}$$

or, what is the same thing,

$$x = \mathrm{A} + \mathrm{R}\,\mathrm{f},$$
$$y = \mathrm{B} + \mathrm{R}\,\mathrm{g},$$
$$z = \mathrm{C} + \mathrm{R}\,\mathrm{h},$$

and if in the foregoing formulæ the point Z′ is taken to be the orbit-pole (longitude $b = \mathrm{G} + 90°$, and colatitude $c = \cos^{-1} \cos \mathrm{N} \cos \mathrm{H}$ as above) then the equation of the orbit-plane is $z' = 0$. We have therefore merely to transform the equations of the ray to the new axes by writing for x, y, z, the values

$$\alpha x' + \alpha' y' + \alpha'' z',$$
$$\beta x' + \beta' y' + \beta'' z',$$
$$\gamma x' + \gamma' y' + \gamma'' z',$$

and then putting $z' = 0$, we find x', y', the co-ordinates in the orbit-plane of its intersections with the ray.

47. The equations thus become,

$$\alpha x' + \alpha' y' - \mathrm{A} - \mathrm{R}\,\mathrm{f} = 0,$$
$$\beta x' + \beta' y' - \mathrm{B} - \mathrm{R}\,\mathrm{g} = 0,$$
$$\gamma x' + \gamma' y' - \mathrm{C} - \mathrm{R}\,\mathrm{h} = 0,$$

or, what is the same thing, we have

$$x' : y' : \mathrm{R} : 1$$
$$= \begin{vmatrix} 1 & & & \\ \alpha & \alpha' & \mathrm{f} & \mathrm{A} \\ \beta & \beta' & \mathrm{g} & \mathrm{B} \\ \gamma & \gamma' & \mathrm{h} & \mathrm{C} \end{vmatrix} : \begin{vmatrix} & 1 & & \\ \alpha & \alpha' & \mathrm{f} & \mathrm{A} \\ \beta & \beta' & \mathrm{g} & \mathrm{B} \\ \gamma & \gamma' & \mathrm{h} & \mathrm{C} \end{vmatrix} : \begin{vmatrix} & & -1 & \\ \alpha & \alpha' & \mathrm{f} & \mathrm{A} \\ \beta & \beta' & \mathrm{g} & \mathrm{B} \\ \gamma & \gamma' & \mathrm{h} & \mathrm{C} \end{vmatrix} : \begin{vmatrix} & & & -1 \\ \alpha & \alpha' & \mathrm{f} & \mathrm{A} \\ \beta & \beta' & \mathrm{g} & \mathrm{B} \\ \gamma & \gamma' & \mathrm{h} & \mathrm{C} \end{vmatrix}$$

$$= \begin{vmatrix} \alpha' & \mathrm{f} & \mathrm{A} \\ \beta' & \mathrm{g} & \mathrm{B} \\ \gamma' & \mathrm{h} & \mathrm{C} \end{vmatrix} : - \begin{vmatrix} \alpha & \mathrm{f} & \mathrm{A} \\ \beta & \mathrm{g} & \mathrm{B} \\ \gamma & \mathrm{h} & \mathrm{C} \end{vmatrix} : - \begin{vmatrix} \mathrm{A} & \alpha & \alpha' \\ \mathrm{B} & \beta & \beta' \\ \mathrm{C} & \gamma & \gamma' \end{vmatrix} : \begin{vmatrix} \mathrm{f} & \alpha & \alpha' \\ \mathrm{g} & \beta & \beta' \\ \mathrm{h} & \gamma & \gamma' \end{vmatrix}.$$

In these formulæ we have identically

$$\beta\gamma' - \beta'\gamma,\ \gamma\alpha' - \gamma'\alpha,\ \alpha\beta' - \alpha'\beta = \alpha'', \beta'', \gamma''$$

and if we write moreover

$$\mathrm{a}, \mathrm{b}, \mathrm{c}, = \mathrm{C}\mathrm{g} - \mathrm{B}\mathrm{h},\ \mathrm{A}\mathrm{h} - \mathrm{C}\mathrm{f},\ \mathrm{B}\mathrm{f} - \mathrm{A}\mathrm{g},$$

(whence identically

$$af + bg + ch = 0$$

and where (a, b, c, f, g, h) are the "six co-ordinates" of the ray), then we have the very simple formulæ

$$\begin{aligned} x' : y' : R : 1 \\ = (a, b, c)(\alpha', \beta', \gamma') : -(a, b, c)(\alpha, \beta, \gamma) : (A, B, C)(\alpha'', \beta'', \gamma'') : (f, g, h)(\alpha'', \beta'', \gamma''),\end{aligned}$$

or omitting (as not required for the present purpose) one of the proportional terms, we have

$$x' : y' : 1 = (a, b, c)(\alpha', \beta', \gamma') : -(a, b, c)(\alpha, \beta, \gamma) : (f, g, h)(\alpha'', \beta'', \gamma''),$$

which are the required expressions for the co-ordinates.

48. Consider in the equations just obtained the axis of x' as fixed but H as variable; that is, let the orbit-pole Z' describe a great circle about the fixed pole X' (longitude G, co-latitude $90° + N$). We have x', y', 1, proportional to linear functions of sin H, cos H; viz., writing for shortness,

$$\begin{aligned} X_c &= \quad -a \sin G + b \cos G, \\ X_s &= (-a \cos G - b \sin G) \sin N - c \cos N, \\ Y_0 &= (-a \cos G - b \sin G) \cos N + c \sin N, \\ W_c &= \quad (f \cos G + g \sin G) \sin N + h \cos N, \\ W_s &= (-f \sin G + g \cos G), \end{aligned}$$

we have,

$$x' = \frac{X_c \cos H + X_s \sin H}{W_c \cos H + W_s \sin H},$$

$$y' = \frac{Y_0}{W_c \cos H + W_s \sin H}.$$

49. I write

$$\left\{ \begin{aligned} &\frac{W_c}{Y_0} = \frac{l}{m} \cos \Delta, \qquad \frac{W_s}{Y_0} = \frac{l}{m} \sin \Delta, \\ &\frac{X_c}{Y_0} = \frac{l}{m} \cos \Delta - \cot J \sin \Delta, \\ &\frac{X_s}{Y_0} = \frac{l}{m} \sin \Delta + \cot J \cos \Delta, \end{aligned} \right.$$

equations which determine m, Δ, l, δ, viz., we have

$$\tan\Delta = \frac{W_y}{W_x},\quad m = \frac{Y_z}{\sqrt{W_x^2 + W_y^2}},$$

$$l = m\,\frac{X_x\cos\Delta + X_y\sin\Delta}{Y_z} = \frac{1}{W_x^2 + W_y^2}\,(X_x W_x + X_y W_y),$$

$$\cot\delta = \frac{-X_x\sin\Delta + X_y\cos\Delta}{Y_z} = \frac{1}{Y_z\sqrt{W_x^2 + W_y^2}}\,(X_y W_x - X_x W_y),$$

and we then very easily find

$$\begin{aligned} x' &= l + m\cot\delta\tan(H - \Delta), \\ y' &= \phantom{l + {}} m\sec(H - \Delta), \end{aligned}$$

and thence also

$$y'^2 - (x' - l)^2\tan^2\delta = m^2;$$

viz. the orbit-plane revolving about the fixed axis S X′, meets the ray in a series of points forming in the orbit-plane a hyperbola having the line S X′ for its conjugate axis.

50. As already remarked (*ante*, No. 11), this hyperbola is nothing else than the intersection of the orbit-plane regarded as fixed, by the hyperboloid generated by the rotation of the ray about the axis S X′. And we thus see the interpretation of the constants, viz.

l is the distance from S along the axis S X′ of the "arm," or shortest distance of S X′ and the ray.

m is the length of this arm.

δ is the inclination of the ray to the axis S X′.

And for the remaining quantity Δ, imagine (through S) parallel to the ray a line through S meeting the sphere in L (L is the pole of the separator), I say that $\Delta - H$ is the angle L X′ Z′: or (what is the same thing) drawing X′ L to meet Π Z′ Y′ in Λ, we have $\Pi\Lambda = \Delta = H + Z'\Lambda$, or (what is the same thing) $Z'\Lambda = \Delta - H$.

51. To verify this, observe that the cosine distances of L from X, Y, Z, are as f : g : h; and thence its cosine distances from X′, Y′, Z′, are as (f, g, h)(α, β, γ) : (f, g, h)$(\alpha', \beta', \gamma')$: (f, g, h)$(\alpha'', \beta'', \gamma'')$; say, for a moment, as f′ : g′ : h′.

Now L A is the perpendicular from L on the side Y′ Z′ of the quadrantal spherical triangle L Y′ Z′, and we thence have

$$\frac{h'}{g'} = \frac{\cos A Y'}{\cos A Z'} = \tan A Z' = \tan (\Delta - H),$$

if Δ has the geometrical signification just assigned to it. But this equation is

$$g' \cos (H - \Delta) + h' \sin (H - \Delta) = 0,$$

that is

$$\tan \Delta = \frac{g' \cos H + h' \sin H}{- g' \sin H + h' \cos H},$$

or substituting for g′, h′ their values, the numerator is

$$f (\alpha' \cos H + \alpha'' \sin H) + g (\beta' \cos H + \beta'' \sin H) + h (\gamma' \cos H + \gamma'' \sin H),$$

which is

$$= - f \sin G + g \cos G, = W_y,$$

and the denominator is

$$f (- \alpha' \sin H + \alpha'' \cos H) + g (- \beta' \sin H + \beta'' \cos H) + h (- \gamma' \sin H + \gamma'' \cos H),$$

which is

$$= (f \cos G + g \sin G) \sin N + h \cos N, = W_x,$$

so that the formula becomes

$$\tan \Delta = \frac{W_y}{W_x},$$

which is the original expression of tan Δ.

52. We might in the equations

$$x' : y' : 1 = (a, b, c)(\alpha', \beta', \gamma') : -(a, b, c)(\alpha, \beta, \gamma) : (f, g, h)(\alpha'', \beta'', \gamma'')$$

consider for instance G or N as alone variable, and then eliminate the variable parameter so as to obtain a locus; but the results would be complicated and the geometrical interpretations not very obvious.

53. I assume (as was done before) N = 0, G = *b* − 90°, H = *c*, that is, the position of the orbit-pole Z′ is longitude *b*, colatitude *c*, and the axis S X′ is the line of nodes or intersection of the orbit-plane with the ecliptic, viz., the longitude of this line is = *b* − 90°.

The formulæ become

$$x' : y' : 1 = (\mathfrak{a} \cos b + \mathfrak{b} \sin b) \cos c - \mathfrak{c} \sin c$$
$$: - \mathfrak{a} \sin b + \mathfrak{b} \cos b$$
$$: (\mathfrak{f} \cos b + \mathfrak{g} \sin b) \sin c + \mathfrak{h} \cos c,$$

or if these are

$$x' = \frac{X_1 \cos c + X_3 \sin c}{W_1 \cos c + W_3 \sin c},$$

$$y' = \frac{Y_2}{W_1 \cos c + W_3 \sin c},$$

the values now are

$$X_1 = \mathfrak{a} \cos b + \mathfrak{b} \sin b$$
$$X_3 = - \mathfrak{c}$$
$$Y_2 = - \mathfrak{a} \sin b + \mathfrak{b} \cos b$$
$$W_1 = \mathfrak{h}$$
$$W_3 = \mathfrak{f} \cos b + \mathfrak{g} \sin b$$

and thence forming as before the values of tan Δ, l, m, cot ϑ, and putting for shortness

$$\sqrt{W_1^2 + W_3^2}, = \sqrt{\mathfrak{h}^2 + (\mathfrak{f} \cos b + \mathfrak{g} \sin b)^2}, = \Omega$$

we find after some easy reductions

$$\tan \Delta = \frac{\mathfrak{f}}{\mathfrak{h}} \cos b + \frac{\mathfrak{g}}{\mathfrak{h}} \sin b,$$

$$m = \frac{1}{\Omega} (-\mathfrak{a} \sin b + \mathfrak{b} \cos b),$$

$$l = \frac{1}{\Omega^2} [(\mathfrak{a}\mathfrak{h} - \mathfrak{c}\mathfrak{f}) \cos b + (\mathfrak{b}\mathfrak{h} - \mathfrak{c}\mathfrak{g}) \sin b],$$

$$\cot \vartheta = \frac{1}{\Omega Y_2} (- \mathfrak{a} \sin b + \mathfrak{b} \cos b) (- \mathfrak{f} \sin b + \mathfrak{g} \cos b),$$

$$= \frac{1}{\Omega} (- \mathfrak{f} \sin b + \mathfrak{g} \cos b),$$

and with these values

$$x' = l + m \cot \vartheta \tan (c - \Delta),$$
$$y' = m \sec (c - \Delta).$$

and thence

$$y'^2 - (x' - l)^2 \tan^2 \vartheta = m^2,$$

viz., this is the hyperbola obtained by rotating the orbit-plane about the line of nodes, longitude $\delta - 90^\circ$.

54. Imagine the orbit-plane (having upon it the hyperbola) brought by such rotation into the plane $z = 0$, or plane of the ecliptic, so that the hyperbola will be a curve in this plane, the inclination to Sx, or longitude of the axis Sx', being of course $= \delta - 90^\circ$. Transforming the equation to axes Sx, Sy, we must write in the equation

$$x' = x \sin \delta - y \cos \delta,$$
$$y' = x \cos \delta + y \sin \delta,$$

and the equation thus becomes

$$(x \cos \delta + y \sin \delta)^2 - (x \sin \delta - y \cos \delta - l)^2 \tan^2 \vartheta = m^2.$$

55. It will be recollected that the equations of the ray were

$$\frac{x - \mathrm{A}}{\mathrm{f}} = \frac{y - \mathrm{B}}{\mathrm{g}} = \frac{z - \mathrm{C}}{\mathrm{h}};$$

writing herein $z = 0$ we find

$$x = \mathrm{A} - \frac{\mathrm{f}}{\mathrm{h}}\mathrm{C}, = \frac{\mathrm{b}}{\mathrm{h}},$$

$$y = \mathrm{B} - \frac{\mathrm{f}}{\mathrm{g}}\mathrm{C}, = -\frac{\mathrm{a}}{\mathrm{h}},$$

and it is clear that this point $\left(\frac{\mathrm{b}}{\mathrm{h}}, -\frac{\mathrm{a}}{\mathrm{h}}\right)$ should lie on the hyperbola.

Substituting for (x, y) the values in question, we have first

$$\mathrm{b} \sin \delta + \mathrm{a} \cos \delta - \mathrm{h}\, l$$

$$= \frac{1}{\Omega^2} \Big\{ \left(\mathrm{h}^2 + (\mathrm{f} \cos \delta + \mathrm{g} \sin \delta)^2\right) (\mathrm{b} \sin \delta + \mathrm{a} \cos \delta)$$

$$- \mathrm{h}\,(\mathrm{a\,h} - \mathrm{c\,f})\left(\cos \delta + (\mathrm{b\,h} - \mathrm{c\,g}) \sin \delta\right) \Big\}$$

$$= \frac{1}{\Omega^2}\left\{(\mathrm{f}\cos\theta + \mathrm{g}\sin\theta)^2(\mathrm{b}\sin\theta + \mathrm{a}\cos\theta) + (f\cos\theta + \mathrm{g}\sin\theta)\,\mathrm{ch})\right\}$$

$$= \frac{1}{\Omega^2}(\mathrm{f}\cos\theta + \mathrm{g}\sin\theta)\left\{(\mathrm{f}\cos\theta + \mathrm{g}\sin\theta)(\mathrm{b}\sin\theta + \mathrm{a}\cos\theta) + \mathrm{ch}\,(\cos^2\theta + \sin^2\theta)\right\}$$

$$= \frac{1}{\Omega^2}(\mathrm{f}\cos\theta + \mathrm{g}\sin\theta)(-\mathrm{a}\sin\theta + \mathrm{b}\cos\theta)(\mathrm{f}\sin\theta - \mathrm{g}\cos\theta);$$

or observing that

$$\tan\delta = \frac{-\Omega}{\mathrm{f}\sin\theta - \mathrm{g}\cos\theta},$$

we have

$$(\mathrm{b}\sin\theta + \mathrm{a}\cos\theta - \mathrm{h}l)\tan\delta = -\frac{1}{\Omega}(\mathrm{f}\cos\theta + \mathrm{g}\sin\theta)(-\mathrm{a}\sin\theta + \mathrm{b}\cos\theta);$$

and hence the result of the substitution is at once found to be,

$$(-\mathrm{a}\sin\theta + \mathrm{b}\cos\theta)^2 - \frac{1}{\Omega^2}(-\mathrm{a}\sin\theta + \mathrm{b}\cos\theta)^2(\mathrm{g}\sin\theta + \mathrm{f}\cos\theta)^2 = \omega^2\mathrm{h}^2$$

$$= \frac{\mathrm{h}^2(-\mathrm{a}\sin\theta + \mathrm{b}\cos\theta)^2}{\Omega^2};$$

viz., the factor $(-\mathrm{a}\sin\theta + \mathrm{b}\cos\theta)^2$ divides out, and the equation then becomes

$$1 - \frac{1}{\Omega^2}(\mathrm{g}\sin\theta + \mathrm{f}\cos\theta)^2 = \frac{\mathrm{h}^2}{\Omega^2};$$

that is

$$\Omega^2 = \mathrm{h}^2 + (\mathrm{g}\sin\theta + \mathrm{f}\cos\theta)^2,$$

which is in fact the value of Ω^2.

56. I seek for the direction of the hyperbola at the point $\left(\frac{\mathrm{b}}{\mathrm{h}}, -\frac{\mathrm{a}}{\mathrm{h}}\right)$ in question. We have

$$dx : dy = (\mathrm{b}\cos\theta - \mathrm{a}\sin\theta)\sin\theta + \cos\theta\tan^2\delta\,(\mathrm{b}\sin\theta + \mathrm{a}\cos\theta - \mathrm{h}l)$$
$$: -(b\cos\theta - \mathrm{a}\sin\theta)\cos\theta + \sin\theta\tan^2\delta\,(\mathrm{b}\sin\theta + \mathrm{a}\cos\theta - \mathrm{h}l),$$

and from the above values of $(\mathrm{b}\sin\theta + \mathrm{a}\cos\theta - \mathrm{h}l)$ and $\tan\delta$, we have

$$\tan^2\delta\,(\mathrm{b}\sin\theta + \mathrm{a}\cos\theta - \mathrm{h}l) = \frac{\mathrm{g}\sin\theta + \mathrm{f}\cos\theta}{\mathrm{f}\sin\theta - \mathrm{g}\cos\theta}(-\mathrm{a}\sin\theta + \mathrm{b}\cos\theta);$$

whence

$$dx : dy = (\mathrm{b}\cos\delta - \mathrm{a}\sin\delta)\sin\delta\,(\mathrm{f}\sin\delta - \mathrm{g}\cos\delta) + (\mathrm{g}\sin\delta + \mathrm{f}\cos\delta)\cos\delta\,(-\mathrm{a}\sin\delta + \mathrm{b}\cos\delta)$$
$$: -(\mathrm{b}\cos\delta - \mathrm{a}\sin\delta)\cos\delta\,(\mathrm{f}\sin\delta - \mathrm{g}\cos\delta) + (\mathrm{g}\sin\delta + \mathrm{f}\cos\delta)\sin\delta\,(-\mathrm{a}\sin\delta + \mathrm{b}\cos\delta),$$

which, multiplying out and reducing by means of the relation $\mathrm{af} + \mathrm{bg} + \mathrm{ch} = 0$, becomes

$$dx : dy = (-\mathrm{a}\sin\delta + \mathrm{b}\cos\delta)(\sin^2\delta + \cos^2\delta)\,\mathrm{f} : (-\mathrm{a}\sin\delta + \mathrm{b}\cos\delta)(\sin^2\delta + \cos^2\delta)\,\mathrm{g};$$

that is

$$dx : dy = \mathrm{f} : \mathrm{g}, \quad \text{or} \quad \frac{dy}{dx} = \frac{\mathrm{g}}{\mathrm{f}},$$

which shows that the hyperbola at the point $\left(\frac{\mathrm{b}}{\mathrm{h}}, -\frac{\mathrm{a}}{\mathrm{h}}\right)$, where it meets the ray, touches the projection

$$\frac{x - \mathrm{A}}{\mathrm{f}} = \frac{y - \mathrm{B}}{\mathrm{g}}$$

of the ray on the plane of xy, which contains the hyperbola.

57. We may consider various particular forms of the hyperbola $y'^2 - (x' - l)^2 \tan^2\delta = m^2$.

1°. If $\tan\delta = 0$, the hyperbola is the pair of parallel lines $y'^2 = m^2$.

This can only happen if $\mathrm{h} = 0$, $\mathrm{f}\cos\delta + \mathrm{g}\sin\delta = 0$. The first equation gives $\mathrm{af} + \mathrm{bg} = 0$, whence $\tan\delta = -\frac{\mathrm{f}}{\mathrm{g}} = \frac{\mathrm{b}}{\mathrm{a}}$; we have thus $m = \frac{-\mathrm{a}\sin\delta + \mathrm{b}\cos\delta}{\mathrm{h}} = \frac{0}{0}$, which is consistent with m finite. The equations show that the ray is parallel to the line of nodes.

2°. If $\tan\delta = \infty$, the hyperbola is $(x' - l)^2 = 0$, viz., the line $x' = l$ twice: the condition is $-\mathrm{f}\sin b + \mathrm{g}\cos\delta = 0$; viz., the ray (not in general cutting the line of nodes) is at right angles to the line of nodes.

3°. If $m = 0$, the hyperbola is the pair of intersecting lines $y'^2 = (x' - l)^2 \tan^2\delta$. The condition is $-\mathrm{a}\sin\delta + \mathrm{b}\cos\delta = 0$, signifying that the ray cuts the line of nodes.

4°. We may have simultaneously $\tan\delta = \infty$, $m = 0$. The hyperbola (as in 2°) is $(x' - l)^2 = 0$. The conditions are $-\mathrm{f}\sin\delta + \mathrm{g}\cos\delta = 0$, $-\mathrm{a}\sin\delta + \mathrm{b}\cos\delta = 0$, whence $\tan\delta = \frac{\mathrm{g}}{\mathrm{f}} = \frac{\mathrm{b}}{\mathrm{a}}$, and therefore also $\mathrm{ag} - \mathrm{bf} = 0$; these signify that the ray cuts at right angles the line of nodes.

The line $x' = l$ passes through the point $\left(\frac{\mathbf{b}}{\mathbf{h}}, -\frac{\mathbf{a}}{\mathbf{h}}\right)$, that is, we ought to have $\mathbf{h}^2 l^2 = \mathbf{a}^2 + \mathbf{b}^2$. The value of l is in the first instance given in the form

$$l = \frac{1}{\Omega^2}\{(\mathbf{a}\mathbf{h} - \mathbf{c}\mathbf{f}) \cos b + (\mathbf{b}\mathbf{h} - \mathbf{c}\mathbf{g}) \sin b\},$$

where

$$\Omega^2 = \mathbf{h}^2 + (\mathbf{f} \cos b + \mathbf{g} \sin b)^2 = \mathbf{h}^2 + \mathbf{f}^2 + \mathbf{g}^2 - (-\mathbf{f} \sin b + \mathbf{g} \cos b)^2 = \mathbf{f}^2 + \mathbf{g}^2 + \mathbf{h}^2.$$

But observe that the equations

$$\mathbf{a}\mathbf{g} - \mathbf{b}\mathbf{f} = 0,$$
$$\mathbf{b}\mathbf{g} + \mathbf{a}\mathbf{f} = -\mathbf{c}\mathbf{h},$$

give

$$\mathbf{f} = \frac{-\mathbf{c}\mathbf{h}}{\mathbf{a}^2 + \mathbf{b}^2}\mathbf{a}, \quad \mathbf{g} = \frac{-\mathbf{c}\mathbf{h}}{\mathbf{a}^2 + \mathbf{b}^2}\mathbf{b},$$

and thence

$$\Omega^2 = \mathbf{f}^2 + \mathbf{g}^2 + \mathbf{h}^2 = \mathbf{h}^2\left(1 + \frac{\mathbf{c}^2}{\mathbf{a}^2 + \mathbf{b}^2}\right) = \frac{\mathbf{h}^2(\mathbf{a}^2 + \mathbf{b}^2 + \mathbf{c}^2)}{\mathbf{a}^2 + \mathbf{b}^2},$$

$$\mathbf{a}\mathbf{h} - \mathbf{c}\mathbf{f} = \mathbf{a}\mathbf{h}\frac{\mathbf{a}^2 + \mathbf{b}^2 + \mathbf{c}^2}{\mathbf{a}^2 + \mathbf{b}^2} = \frac{\mathbf{a}}{\mathbf{h}}\Omega,$$

$$\mathbf{b}\mathbf{h} - \mathbf{c}\mathbf{g} = \mathbf{b}\mathbf{h}\frac{\mathbf{a}^2 + \mathbf{b}^2 + \mathbf{c}^2}{\mathbf{a}^2 + \mathbf{b}^2} = \frac{\mathbf{b}}{\mathbf{h}}\Omega;$$

consequently

$$l = \frac{1}{\Omega^2 \mathbf{h}}(\mathbf{a} \cos b + \mathbf{b} \sin b)\,\Omega^2 = \frac{1}{\mathbf{h}}(\mathbf{a} \cos b + \mathbf{b} \sin b) = \frac{1}{\mathbf{h}}\sqrt{\mathbf{a}^2 + \mathbf{b}^2},$$

which is right.

58. I return to the equation of the hyperbola written in the form

$$(x \cos b + y \sin b)^2 - (x \sin b - y \cos b - l)^2 \tan^2 i = m^2;$$

being (as was shown) a hyperbola passing through the point $\left(\frac{\mathbf{b}}{\mathbf{h}}, -\frac{\mathbf{a}}{\mathbf{h}}\right)$ where its plane is met by the ray, and touching at this point the projection $\frac{x - A}{\mathbf{f}} = \frac{y - B}{\mathbf{g}}$.

If in the equation we consider b as variable, we have a series of hyperbolas, viz., these are the intersections of the plane of xy with the hyperboloids of revolution obtained by making the ray rotate successively round the several lines $x \cos b + y \sin b = 0$ through the focus S. And, as just seen, these hyperbolas all of them touch at $\left(\frac{\mathrm{b}}{\mathrm{h}}, -\frac{\mathrm{a}}{\mathrm{h}}\right)$ the projection of the ray.

59. The hyperbola to any particular angle b is the hyperbola belonging to the ray, in the planogram for an orbit-plane rotating about the axis $x \cos b + y \sin b = 0$; so that the system of hyperbolas would be useful for the construction of any such planogram. And there is another series of curves which, if they could be constructed with moderate facility, would be very useful for the same purpose; viz., reverting to the equations

$$\begin{aligned} x' : y' : 1 = & (\mathrm{a} \cos b + \mathrm{b} \sin b) \cos c - \mathrm{c} \sin c \\ : & \ -\mathrm{a} \sin b + \mathrm{b} \cos b \\ : & \ (\mathrm{f} \cos b + \mathrm{g} \sin b) \sin c + \mathrm{h} \cos c, \end{aligned}$$

which determine in the orbit-plane the co-ordinates x', y', of the intersection thereof with the ray: imagine as before that the point is marked on the orbit-plane, and let it by a rotation of the orbit-plane be brought into the plane of xy; so that x', y', will be the co-ordinates in the direction of and perpendicular to the line of nodes of a point on the hyperbola $y'^2 - (x' - l)^2 \tan^2 \delta = m^2$, or $(x \cos b + y \sin b)^2 - (x \sin b - y \cos b - l)^2 \tan^2 \delta = m^2$, viz., of the point corresponding to an orbit-pole, colatitude c. Suppose that x, y, are the co-ordinates of this same point referred to the fixed axes, we have

$$\begin{aligned} x &= x' \sin b + y' \cos b, \\ y &= -x' \cos b + y' \sin b, \end{aligned}$$

and thence

$$\begin{aligned} x : y : 1 = & \ (\mathrm{a} \cos b + \mathrm{b} \sin b) \sin b \cos c - \mathrm{c} \sin b \sin c + (-\mathrm{a} \sin b + \mathrm{b} \cos b) \cos b \\ : & -(\mathrm{a} \cos b + \mathrm{b} \sin b) \cos b \cos c + \mathrm{c} \cos b \sin c + (-\mathrm{a} \sin b + \mathrm{b} \cos b) \sin b \\ : & \ (\mathrm{f} \cos b + \mathrm{g} \sin b) \sin c + \mathrm{h} \cos c, \end{aligned}$$

the co-ordinates of the point just referred to. Now, if from these equations we could eliminate b, we should have a series of curves containing the variable parameter c, intersecting the series of hyperbolas; and thus marking out on each of these hyperbolas the points which belong to the successive values of the parameter c; we should thus have in the plane of xy the point corresponding to an orbit-pole longitude b and colatitude c. The series of curves in question may be called "graduation curves," viz., they would serve for the graduation of the hyperbola in the planogram for an orbit-plane rotating round any line $x \cos b + y \sin b = 0$ in the plane of xy. But the elimination cannot be easily effected, and I am not in possession of any method of tracing the series of curves.

60. I remark that from the equations

$$\begin{aligned} x' : y' : 1 = \; & (\mathrm{a} \cos b + \mathrm{h} \sin b) \cos c - \mathrm{c} \sin c \\ : & - \mathrm{a} \sin b + \mathrm{b} \cos b \\ : & \; (\mathrm{f} \cos b + \mathrm{g} \sin b) \sin c + \mathrm{h} \cos c, \end{aligned}$$

we may without difficulty eliminate b; the result is, in fact,

$$\begin{aligned} & [x' (-\mathrm{a\,h} \cos c) + y' (-\mathrm{b\,h} \cos^2 c - \mathrm{c\,g} \sin^2 c) - \mathrm{a\,c} \sin c]^2 \\ + & [x' (\;\;\mathrm{b\,h} \cos c) + y' (-\mathrm{a\,h} \cos^2 c + \mathrm{c\,f} \sin^2 c) + \mathrm{b\,c} \sin c]^2 \\ = & [x' (\;\;\mathrm{c\,h} \sin c) + y' (\;\;\mathrm{a\,g} - \mathrm{b\,f}) \sin c \cos c + (\mathrm{a}^2 - \mathrm{b}^2) \cos c]^2, \end{aligned}$$

a conic; but the geometrical signification of this result is not obvious, and I do not make any use of it.

Article Nos. 61 to 63. The Trivector and the Orbit.

61. Considering now the three rays, these are determined by their six co-ordinates,

$$\begin{aligned} & (\mathrm{a}_1, \mathrm{b}_1, \mathrm{c}_1, \mathrm{f}_1, \mathrm{g}_1, \mathrm{h}_1), \\ & (\mathrm{a}_2, \mathrm{b}_2, \mathrm{c}_2, \mathrm{f}_2, \mathrm{g}_2, \mathrm{h}_2), \\ & (\mathrm{a}_3, \mathrm{b}_3, \mathrm{c}_3, \mathrm{f}_3, \mathrm{g}_3, \mathrm{h}_3), \end{aligned}$$

respectively; and the intersections with the orbit-plane are given by

$$\begin{aligned} x'_1 : y'_1 : 1 &= (\mathrm{a}_1, \mathrm{b}_1, \mathrm{c}_1)(\alpha, \beta, \gamma) : -(\mathrm{a}_1, \mathrm{b}_1, \mathrm{c}_1)(\alpha', \beta', \gamma') : (\mathrm{f}_1, \mathrm{g}_1, \mathrm{h}_1)(\alpha'', \beta'', \gamma''), \\ x'_2 : y'_2 : 1 &= (\mathrm{a}_2, \mathrm{b}_2, \mathrm{c}_2)(\;\;,,\;\;) : -(\mathrm{a}_2, \mathrm{b}_2, \mathrm{c}_2)(\;\;,,\;\;) : (\mathrm{f}_2, \mathrm{g}_2, \mathrm{h}_2)(\;\;,,\;\;), \\ x'_3 : y'_3 : 1 &= (\mathrm{a}_3, \mathrm{b}_3, \mathrm{c}_3)(\;\;,,\;\;) : -(\mathrm{a}_3, \mathrm{b}_3, \mathrm{c}_3)(\;\;,,\;\;) : (\mathrm{f}_3, \mathrm{g}_3, \mathrm{h}_3)(\;\;,,\;\;), \end{aligned}$$

where the axes $\mathrm{S}x'$, $\mathrm{S}y'$, are an arbitrary set of rectangular axes in the orbit-plane; or where, as before, the axis Sx' may be taken to be the line of nodes.

There is no difficulty in finding the equation of the orbit. Writing $r_1 = \sqrt{x'^2_1 + y'^2_1}$, we have

$$r_1 = \frac{u_1}{(f_1, g_1, h_1)(\alpha'', \beta'', \gamma'')}$$

if

$$u_1 = \pm \sqrt{[(a_1, b_1, c_1)(\alpha', \beta', \gamma')]^2 + [(a_1, b_1, c_1)(\alpha, \beta, \gamma)]^2},$$

the sign being taken in such manner that r_1 shall be positive; viz., the sign must be the same as that of $(f_1, g_1, h_1)(\alpha'', \beta'', \gamma'')$. And we have the like formulæ for r_2 and r_3. Substituting these values, the equation of the orbit becomes

$$\begin{vmatrix} r, & x', & y', & 1 \\ u_1, & (a_1, b_1, c_1)(\alpha', \beta', \gamma'), & -(a_1, b_1, c_1)(\alpha, \beta, \gamma), & (f_1, g_1, h_1)(\alpha'', \beta'', \gamma'') \\ u_2, & (a_2, b_2, c_2)(\quad " \quad), & -(a_2, b_2, c_2)(\quad " \quad), & (f_2, g_2, h_2)(\quad " \quad) \\ u_3, & (a_3, b_3, c_3)(\quad " \quad), & -(a_3, b_3, c_3)(\quad " \quad), & (f_3, g_3, h_3)(\quad " \quad) \end{vmatrix} = 0.$$

62. Considering the minor determinants formed with the terms under the x' and y', for instance—

$$\begin{matrix} (a_1, b_1, c_1)(\alpha', \beta', \gamma'), & -(a_1, b_1, c_1)(\alpha, \beta, \gamma) \\ +(a_2, b_2, c_2)(\alpha, \beta, \gamma), & (a_2, b_2, c_2)(\alpha', \beta', \gamma') \end{matrix}$$

this is

$$\begin{aligned} = \; & (b_1 c_2 - b_2 c_1)(\beta\gamma' - \beta'\gamma) \\ & + (c_1 a_2 - c_2 a_1)(\gamma\alpha' - \gamma'\alpha) \\ & + (a_1 b_2 - a_2 b_1)(\alpha\beta' - \alpha'\beta) \end{aligned}$$

$$= \alpha''(b_1 c_2 - b_2 c_1) + \beta''(c_1 a_2 - c_2 a_1) + \gamma''(a_1 b_2 - a_2 b_1),$$

or, what is the same thing,

$$= (b_1 c_2 - b_2 c_1 \,,\; c_1 a_2 - c_2 a_1 \,,\; a_1 b_2 - a_2 b_1)(\alpha'', \beta'', \gamma'');$$

with the like expressions for the other two minors. And we thus obtain the following developed form of the equation, viz.

$$\{x'(a_1, b_1, c_1)(\alpha, \beta, \gamma) + y'(a_1, b_1, c_1)(\alpha', \beta', \gamma')\}[-u_2(f_3, g_3, h_3)(\alpha'', \beta'', \gamma'') \\ + u_3(f_2, g_2, h_2)(\alpha'', \beta'', \gamma'')]$$

$$+ \{x'(a_2, b_2, c_2)(\quad " \quad) + y'(a_2, b_2, c_2)(\quad " \quad)\}[-u_3(f_1, g_1, h_1)(\quad " \quad) \\ + u_1(f_3, g_3, h_3)(\quad " \quad)]$$

$$+\{x'(a_3, b_3, c_3)(\alpha, \beta, \gamma) + y'(a_3, b_3, c_3)(\alpha', \beta', \gamma')\}[-u_1(f_2, g_2, h_2)(\alpha'', \beta'', \gamma'') \\ + u_2(f_1, g_1, h_1)(\alpha'', \beta'', \gamma'')]$$

$$+ (b_2 c_3 - b_3 c_2,\ c_2 a_3 - c_3 a_2,\ a_2 b_3 - a_3 b_2)(\alpha'', \beta'', \gamma'')[r(f_1, g_1, h_1)(\alpha'', \beta'', \gamma'') - u_1]$$

$$+ (b_3 c_1 - b_1 c_3,\ c_3 a_1 - c_1 a_3,\ a_3 b_1 - a_1 b_3)(\quad " \quad)[r(f_2, g_2, h_2)(\quad " \quad) - u_2]$$

$$+ (b_1 c_2 - b_2 c_1,\ c_1 a_2 - c_2 a_1,\ a_1 b_2 - a_2 b_1)(\quad " \quad)[r(f_3, g_3, h_3)(\quad " \quad) - u_3] = 0,$$

being an equation of the form $\Omega r = A x' + B y' + C$.

63. The coefficient of r is a quadric function of $(\alpha'', \beta'', \gamma'')$, and if this vanish the orbit is a right line. It thus appears that the orbit will be a right line provided only the orbit-axis be situate in a certain quadric cone, or (what is the same thing) the orbit-pole be situate in a certain spherical conic: agreeing with a preceding result, viz. the cone is that reciprocal to the cone, vertex S, circumscribed about the hyperboloid which contains the three rays. And we see that the equation of this reciprocal cone is

$$\begin{vmatrix} & \alpha'', & \beta'', & \gamma'' \\ (f_1, g_1, h_1)(\alpha'', \beta'', \gamma''), & a_1, & b_1, & c_1 \\ (f_2, g_2, h_2)(\quad " \quad), & a_2, & b_2, & c_2 \\ (f_3, g_3, h_3)(\quad " \quad), & a_3, & b_3, & c_3 \end{vmatrix} = 0.$$

Article Nos. 64 and 65. The Special Symmetrical System of three Rays.

64. In what follows I consider the three rays forming a symmetrical system as already referred to: viz. the three rays intersect the plane of the ecliptic at points equi-distant from S at longitudes 0°, 120°, 240°; each of them is at right angles to the line joining S with the intersection with the plane of the ecliptic, and at an inclination = 60° to this plane: the figure shows the projection on the plane of the ecliptic of the portions which lie above this plane of the three rays respectively.

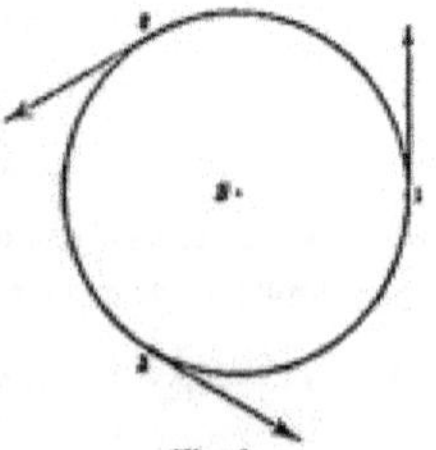

Fig. 6.

The three rays lie on a hyperboloid of revolution having the line Sz for its axis; the circumscribed or asymptotic cone vertex S, is a right cone of the semi-aperture = 30°; the reciprocal cone is therefore a right cone semi-aperture 60°, or (what is the same thing) the regulator is a small circle, angular radius 60°, and the regulator and separators have the positions shown in figure 1, see p. 21.

Taking S 1 = S 2 = S 3 = 1, and writing down the equations of the three rays in the forms

$$\frac{x-1}{0} = \frac{y}{1} = \frac{z}{\tan 60^\circ},$$

$$\frac{x+\cos 60^\circ}{-\sin 60^\circ} = \frac{y-\sin 60^\circ}{-\cos 60^\circ} = \frac{z}{\tan 60^\circ},$$

$$\frac{x+\cos 60^\circ}{\sin 60^\circ} = \frac{y+\sin 60^\circ}{-\cos 60^\circ} = \frac{z}{\tan 60^\circ},$$

we obtain the six co-ordinates of the three rays respectively

$$(a_1, b_1, c_1, f_1, g_1, h_1) = (\ \ 0, \sqrt{3}, -1, \quad 0, 1, \quad \sqrt{3}),$$

$$(a_2, b_2, c_2, f_2, g_2, h_2) = (\ \ 3, \sqrt{3}, \quad 2, \quad \sqrt{3}, 1, -2\sqrt{3}),$$

$$(a_3, b_3, c_3, f_3, g_3, h_3) = (-3, \sqrt{3}, \quad 2, -\sqrt{3}, 1, -2\sqrt{3}),$$

whence the intersections with the orbit-plane are given by

$$x'_1 : y'_1 : 1 = \beta'\sqrt{3} - \gamma' : -\beta\sqrt{3} + \gamma : \beta'' + \gamma''\sqrt{3},$$

$$x'_2 : y'_2 : 1 = 3\alpha' + \beta'\sqrt{3} + 2\gamma' : -3\alpha - \beta\sqrt{3} - 2\gamma : \alpha''\sqrt{3} + \beta'' - 2\sqrt{3}\gamma'',$$

$$x'_3 : y'_3 : 1 = -3\alpha' + \beta'\sqrt{3} + 2\gamma' : 3\alpha - \beta\sqrt{3} - 2\gamma : -\alpha''\sqrt{3} + \beta'' - 2\sqrt{3}\gamma'',$$

where if (as before) the position of the orbit-plane be determined by means of the longitude b and colatitude c of the orbit-pole, we have

$$\alpha, \beta, \gamma = \sin b, \quad -\cos b, \quad 0,$$

$$\alpha', \beta', \gamma' = \cos b \cos c, \sin b \cos c, -\sin c,$$

$$\alpha'', \beta'', \gamma'' = \cos b \sin c, \sin b \sin c, \quad \cos c,$$

and the passage from the co-ordinates x', y', to x, y, is given by

$$x' = x \sin b - y \cos b,$$

$$y' = x \cos b + y \sin b,$$

or conversely

$$x = \quad x' \sin b + y' \cos b,$$
$$y = - x' \cos b + y' \sin b.$$

65. To develope the results, I consider the orbit-pole as passing through certain series of positions. The locus may be a meridian circle: by reason of the symmetry of the system, the results are not altered by a change of 120° in the longitude of the meridian; so that, by considering the two meridians 0°–180° and 90°–270°, we, in fact, consider twelve half meridians at the intervals of 30°. An illustration is afforded by Plate I.; the orbit-pole describes successively the meridians 0°, 30°, 60°, 90°, and the line 1, by its intersection with the orbit-plane, traces out on this plane a series of hyperbolas shown in the figure; the hyperbola for the meridian 90° is a right line, but (except for the position where the orbit-plane passes through the line 1) the locus is a determinate point on this line. Planogram No. 1 (Plate II.) refers to the meridian 90°–270°, and Planogram No. 2 (Plate III.) to the meridian 0°–180°. Next, if the orbit-pole be at one of the points A, that is, if the orbit-plane pass through a ray—though the position of the orbit-pole be here determinate, yet as there is a series of orbits, this also will give rise to a planogram: I call it Planogram No. 3. The orbit-pole may pass along a separator circle (viz. the orbit-plane be parallel to a ray), this is Planogram No. 4. And, lastly, the orbit-pole may pass along the ecliptic (or the orbit-plane may pass through the axis SZ) I call this Planogram No. 5. But the last three planograms are not considered in the like detail as the first two, and I have not, in regard to them, tabulated the results, nor given any Plates.

Article Nos. 66 to 82. Planogram No. 1, the Meridian 90°–270° (see Plate II).

66. Supposing that the orbit-plane rotates about the axis SZ (fig. 6, see p. 51) in the plane of the ecliptic, the orbit-pole will describe the meridian 90°–270°, the position of the orbit-pole being $b = 90°$, $c = 0°$ to 90°, or else $b = 270°$, $c = 0°$ to 90°. But the same analytical formula extends to the two half meridians, viz. we may take $b = 90°$, and extend c over 180°, in the final results making c an arc between 0° and 90°, and $b = 90°$, or $= 270°$, as the case requires.

67. Assuming then $b = 90°$, we have

$$\begin{aligned} \alpha\,,\ \beta\,,\ \gamma\ &= 1,\quad 0\ ,\quad 0, \\ \alpha',\ \beta',\ \gamma' &= 0,\ \cos c, -\sin c, \\ \alpha'',\ \beta'',\ \gamma'' &= 0,\ \sin c,\ \cos c, \end{aligned}$$

and, moreover, $x', y' = x, y$: so that instead of $(x_1' y_1')$, &c., we may write at once (x_1, y_1), &c. The formulæ become

$$\begin{aligned} x_1 : y_1 : 1 &= \sqrt{3}\cos c + \sin c : \quad 0 : \sin c + \sqrt{3}\cos c, \\ x_2 : y_2 : 1 &= \sqrt{3}\cos c - 2\sin c : -3 : \sin c - 2\sqrt{3}\cos c, \\ x_3 : y_3 : 1 &= \sqrt{3}\cos c - 2\sin c : \quad 3 : \sin c - 2\sqrt{3}\cos c, \end{aligned}$$

that is

$$x_1 = 1,\ y_1 = 0,$$

(viz. the orbit-plane, as is evident, meets the ray 1 in a fixed point, its intersection with the plane of $x\,y$);

$$x_2 = \frac{\sqrt{3}\cos c - 2\sin c}{\sin c - 2\sqrt{3}\cos c}\,,\ x_3 = x_2,$$

$$y_2 = \frac{-3}{\sin c - 2\sqrt{3}\cos c}\,,\ y_3 = -y_2,$$

and writing

$$\frac{2\sqrt{3}}{\sqrt{13}} = \cos\omega,\quad \frac{1}{\sqrt{13}} = \sin\omega,\quad \frac{1}{2\sqrt{3}} = \tan\omega,$$

(whence $\omega = 16° 6'$) we find

$$x_2 = -\frac{8}{13} + \frac{3\sqrt{3}}{13}\tan(c+\omega),$$

$$y_2 = \frac{3}{\sqrt{13}}\sec(c+\omega),$$

and we thence have for the hyperbola, the locus of (x_2, y_2) and (x_3, y_3)

$$\left(x + \frac{8}{13}\right)^2 = \frac{3}{13}\left(y^2 - \frac{9}{13}\right).$$

viz. the points (x_2, y_2) and (x_3, y_3) are situate on the hyperbola, symmetrically on opposite sides of the axis of x. For $c = 0$, we have $x_1 = -\frac{1}{2}$, $y_1 = \frac{\sqrt{3}}{2}$, $(x_1^2 + y_1^2 = 1)$, and the hyperbola at this point touches the circle $x^2 + y^2 = 1$; and similarly for x_{11}, y_{11}. The inclination of the asymptotes to the axis of y is given by $\tan \eta = \sqrt{\frac{3}{13}}$, $\eta = 22° 56'$.

68. The orbits are conics, focus S and vertex 1. It will be convenient to consider c as passing from 0° to 90° − ω, and from 0° to − (90° + ω); that is, from 0° to 73° 54′ − ι, and from 0° to − 116° 6′ + ι, if ι be indefinitely small: the point 2 will thus traverse the upper branch (alone shown in the Plate) of the guide-hyperbola, viz., for c = 0° it will be at the point of contact with the circle; for c = 73° 54′ − ι it will be at ∞, and for c = − 106° 6′ + ι at ∞′. For c = 0° the orbit is the circle; as c increases positively, it becomes an ellipse of increasing excentricity and major axis, until for a certain value (c = 46° 48′ as will appear) it becomes a parabola; it then becomes a hyperbola (concave branch); for c = 52° 45′ it becomes the hyperbola Σ′ subsequently referred to; and for c = 60° (the point 2 being then on the line shown in the figure) the orbit becomes this right line. As c continues to increase, the orbit becomes a hyperbola (convex branch); and ultimately for c = 73° 54′ − ι, the point 2 goes to ∞, and the orbit becomes a hyperbola (convex) Σ, having an asymptote parallel to that of the guide-hyperbola: the inclination to the axis of x being thus 90° − 22° 56′, = 67° 4′.

69. Next as c increases negatively, the point 2 moves from the point of contact in the other direction to ∞′: for c = 0° the orbit is of course the circle, and as c increases negatively the orbits are at first the very same series of orbits as those belonging to the positive values,* viz., they are first ellipses, of increasing excentricity and major axis; then for c = − 92° 54′ the orbit is the parabola; the orbits are then hyperbolas (concave), and finally for c = − 106° 6′ + ι, when 2 is at ∞′, the orbit is a hyperbola Σ′, the asymptote of which is parallel to that of the guide-hyperbola, viz., the inclination to the axis of x is = 67° 4′.

70. It will be observed that the orbits from the circle to the hyperbola Σ′ each intersect the guide-hyperbola (that is, the branch shown in the figure)

* Of course, as corresponding to different values of c, they are not the same orbits in space, but they are only the same curves in the planogram.

in two points, the one corresponding to a positive, the other to a negative value of c; in the positive series, the remaining orbits from the hyperbola Σ', through the right line to the convex hyperbola Σ, each intersect the guide-hyperbola (same branch) in a single point only, for which c is positive.

71. There is, in the passage of the orbit-pole from $c = -106^\circ 6' + \iota$ to $c = 73^\circ 54' - \iota$, say at $c = 73^\circ 54'$, a discontinuity of orbit, viz., an abrupt change from the concave hyperbola Σ' to the concave hyperbola Σ; observe that the direction of the asymptotes being the same in each, the excentricity e has the same value.

The point in question ($b = 90^\circ$, $c = 73^\circ 54'$) is one of the points B of the spherogram, and the hyperbolas Σ, Σ' are two of the four orbits belonging to this point. And, by what precedes, it appears that as the orbit-pole passes through this point along a meridian downwards to the ecliptic the change is from a concave to a convex orbit.

72. On account of the symmetry in regard to the axis of x, the equation of the orbit will be of the form $r = \mathrm{A}\,x + \mathrm{B}$; viz., the equation is at once found to be

$$r - 1 = \frac{r_4 - 1}{x_4 - 1}(x - 1).$$

73. The excentricity is the co-efficient A taken positively ($e = \pm \mathrm{A}$): it is in the present case proper to attend to the value of the co-efficient itself,

$$\mathrm{A} = \frac{r_4 - 1}{x_4 - 1},$$

the sign of A will then indicate the position of the centre of the orbit, viz., according as A is positive or negative the centre will be on the negative or the positive side of the focus S. To investigate the variation of A, we may express it as a function of tan c, $= \lambda$ suppose. We have

$$x_4 = \frac{\sqrt{3 - 2\lambda}}{\lambda - 2\sqrt{3}}, \qquad y_4 = \frac{-3}{\lambda - 2\sqrt{3}}$$

and thence,

$$r_4 = \frac{1}{\lambda - 2\sqrt{3}}\,\mathrm{R}_4, \qquad \mathrm{R}_4 = \pm\sqrt{12 - 4\sqrt{3}\,\lambda + 13\lambda^2}.$$

viz., r_4 must be positive, that is, R_4 is positive or negative according to the sign of $\lambda - 2\sqrt{3}$; negative if $\lambda < 2\sqrt{3}$ or $c < 73^\circ\ 54'$, positive if $\lambda > 2\sqrt{3}$ or $c > 73^\circ\ 54'$. And we have then

$$A = \frac{\lambda - 2\sqrt{3} - R_1}{3(\lambda - \sqrt{3})}.$$

But a more convenient formula is obtained by writing

$$\theta = -\cot c + \frac{1}{2\sqrt{3}},$$

$$\alpha = \qquad\qquad \frac{1}{2\sqrt{3}},$$

we then have

$$\sqrt{1+\theta^2} = \theta r_1,$$

which determines the sign of the radical, viz., this must have the same sign as θ; and then for the co-efficient

$$A = \frac{2}{3(\theta+\alpha)}(-\sqrt{1+\theta^2}+\theta).$$

74. For c a small arc $= \epsilon$, θ is large and negative, and $\sqrt{1+\theta^2}$, having the same sign as θ, is $= \theta + \frac{1}{2\theta}$ nearly; we have therefore

$$A = \frac{2}{3\theta}\cdot\frac{-1}{2\theta} = -\frac{1}{3\theta^2} \text{ approximately.}$$

For c nearly $= 60^\circ$, say $c = 60^\circ \pm \epsilon$,

$$\cot c = \cot 60^\circ \pm \epsilon \operatorname{cosec}^2 60^\circ = \frac{1}{\sqrt{3}} \pm \frac{4\epsilon}{3},$$

$$\theta = -\frac{1}{2\sqrt{3}} \pm \frac{4\epsilon}{3},\quad \theta + \alpha = \pm\frac{4\epsilon}{3},\quad \sqrt{1+\theta^2} = -\frac{\sqrt{13}}{2\sqrt{3}},$$

and thence

$$A = \frac{-1+\sqrt{13}}{2\sqrt{3}} \div \pm\frac{4\epsilon}{3} = \pm\frac{\sqrt{3}}{8}\cdot\frac{-1+\sqrt{13}}{\epsilon},$$

viz., this is $-\infty$ for $c = 60^\circ - \epsilon$, and $+\infty$ for $c = 60^\circ + \epsilon$.

For c nearly $= 90^\circ - \omega$, say first $c = 73^\circ\ 54' - \epsilon$, we have

$$\cot c = \frac{1}{2\sqrt{3}} + \frac{13}{12}\epsilon,\qquad \theta = -\frac{13}{12}\epsilon,\qquad \theta + \alpha = \frac{1}{2\sqrt{3}},\qquad \sqrt{1+\theta^2} = -1.$$

whence

$$A = \frac{4}{\sqrt{3}} = 2{\cdot}30940,$$

But if $c = 73^\circ\ 54' + \epsilon$, then $\theta = \frac{11}{12}\epsilon$, $\theta + \alpha = -\frac{1}{2\sqrt{3}}$, $\sqrt{1+\theta^2} = +1$,

and

$$A = -\frac{4}{\sqrt{3}} = -2{\cdot}30940,$$

viz., there is an abrupt change from $A = +\frac{4}{\sqrt{3}}$ to $A = -\frac{4}{\sqrt{3}}$; corresponding to the discontinuity of orbit already referred to. We may diminish c by 180°, and consider the last-mentioned value, $A = -\frac{4}{\sqrt{3}}$, as belonging to $c = -90^\circ - \omega + \epsilon = -(106^\circ\ 6' - \epsilon)$.

75. Consider next that c passes from 0 to $-(106^\circ\ 6' - \epsilon)$. First if

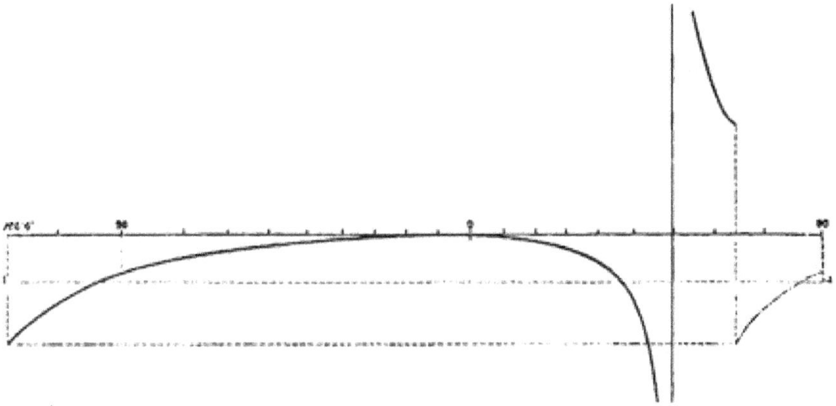

Fig. 7.

c is a small negative quantity $c = -\epsilon$, θ is large and positive, and $\sqrt{1+\theta}$ having the same sign as θ (positive) is $= \theta + \frac{1}{2\theta}$ nearly, we have therefore $A = \frac{2}{3\theta} \cdot \frac{-1}{2\theta} = -\frac{1}{3\theta^2}$ (same as for $c = +\epsilon$). And it is easy to see that as c increases negatively, A is always increasing negatively, its value for $c = -90^\circ$ being $A = \frac{1-\sqrt{13}}{3} = -{\cdot}8685$, and for $c = -106^\circ\ 6'$ $+ 1$ being $= -2{\cdot}30940$ as above. We have a diagram of A (see preceding page).

76. It thus appears that from $A = 0$ to $A = -\frac{4}{\sqrt{3}}$, there are always to any given value of A two values of c or positions of the orbit-pole. In particular if A be $= -1$, the curve will be a parabola; the values of c lying between 0°, 60° and between $73^\circ\ 54'$, 90° respectively.

To find them, writing $A = -1$, we have

$$-3\theta - 3a = 2\theta - 2\sqrt{1+\theta^2},\ \text{that is, } 5\theta + 3a = 2\sqrt{1+\theta^2},$$

or

$$21\theta^2 + 30a\theta + 9a^2 - 4 = 0,$$

that is, substituting for a its value $= \frac{1}{2\sqrt{3}}$,

$$21\theta^2 + 5\sqrt{3}\,\theta - \frac{13}{4} = 0,\ (14\theta\sqrt{3} + 5)^2 = 116,$$

or

$$\theta = \frac{-5 \pm \sqrt{116}}{14\sqrt{3}},$$

that is

$$\theta = -{\cdot}65034, \qquad \theta = {\cdot}13797,$$

giving

$$\cot c = +{\cdot}93902, \qquad \cot c = +{\cdot}05071,$$

or

$$c = 46^\circ\ 48', \qquad c = 87^\circ\ 6'.$$

77. It has been seen that $c = 73^\circ\ 54' + 1$ gives $A = -\frac{4}{\sqrt{3}} = -2{\cdot}30940$; there will be between 0° and 60° another value of c, giving for A this same value; to find this value of c write $A = -\frac{4}{\sqrt{3}}$, then we have

$$-4\sqrt{3}\left(\theta + \frac{1}{2\sqrt{3}}\right) = 2\theta - 2\sqrt{1+\theta^2},$$

that is

$$(1 + 2\sqrt{3})\theta + 1 = \sqrt{1+\theta^2}.$$

or

$$(12 + 4\sqrt{3})\theta^2 + (2 + 4\sqrt{3})\theta = 0,$$

satisfied as it should be by $\theta = 0$, and also by

$$\theta = -\frac{1 + 2\sqrt{3}}{2(3 + \sqrt{3})} = -\cdot47170,$$

giving

$$\cot c = \cdot76038 \quad \text{or } c = 52^\circ\ 45'.$$

78. Representing the equation of the orbit by

$$r = \mathrm{A}x \pm a(1 - \mathrm{A}^2),$$

we have for the point 1,

$$1 = \mathrm{A} \pm a(1 - \mathrm{A}^2),$$

that is

$$a = \frac{\pm 1}{1 + \mathrm{A}},$$

where the sign is to taken so that a shall be positive.

79. With a view to the calculation of the times of passage, I calculate a series of values of x_2, y_2, r_2, A, a, for values of c at the intervals of 5° and for a few intermediate values; we have $x_3, y_3, r_3 = x_2, y_2, r_2$, so that these are known; so long as the orbit is an ellipse, the time of passage between the points 2 and 3, say T_{23}, may be calculated by LAMBERT'S equation, the length of the chord $y_2 - y_3 = 2y_2$ being known without any fresh calculation. And then the times T_{12} and T_{31} being equal, and the sum $\mathrm{T}_{12} + \mathrm{T}_{23} + \mathrm{T}_{31}$ being equal to the whole periodic time $\left(\text{reckoned as} = 3a^{\frac{3}{2}}\right)$ the times T_{12} and T_{31} are also known. But when the orbit is a concave hyperbola there is no time T_{23}, and the other two times $\mathrm{T}_{12} = \mathrm{T}_{31}$ must be calculated. For the reason referred to (*ante*, No. 39) I did not use LAMBERT'S equation,—and it was less necessary to do so, by reason that, the transverse axis coinciding with the axis of x, the other formula could be employed without difficulty.

80. The formulæ for x_2, y_2 adapted to logarithmic calculation are

$$\log(x_2 + \cdot61539) = \overline{11}\cdot60174 + \log\tan(c + 16^\circ\ 6'),$$
$$\log y_2 \qquad\qquad = \overline{11}\cdot92015 + \log\sec(c + 16^\circ\ 6').$$

where y_2 is always positive, but the sign of x_2 must be attended to. The values of r_2 and its inclination ϕ_2 to the axis of x are then to be calculated from

$$\tan\phi_2 = \frac{y_2}{x_2}; \quad r_2 = x_2 \sec\phi_2 \text{ or } = y_2 \operatorname{cosec}\phi_2,$$

(viz. for r_2 it is proper to use the first or the second value, according as x_2 is greater or less than y_2).

We have then $e = (\pm A)$ and a from the foregoing formulæ

$$A = \frac{r_2 - 1}{x_2 - 1}, \quad a = \frac{\pm 1}{1 + A},$$

where a, e are each of them positive.

And then for the Times

$$T_{12} = \frac{3}{2\pi}(\chi - \chi' - \sin\chi + \sin\chi'); \quad \left(\log\frac{3}{2\pi} = \bar{1}\cdot67894\right).$$

where

$$a\cos\chi = a - r_2 - y_2,$$
$$a\cos\chi' = a - r_2 + y_2,$$

and attention is necessary in order to the selection of the proper values of the angles χ, χ'.

And finally

$$T_{23} = T_{31} = \tfrac{1}{2}(3a^{\frac{3}{2}} - T_{12}).$$

81. I subjoin a specimen; the characteristics of the logarithms are (as in the actual calculations) omitted.

$\delta = 90° \quad c = 20°,$
$c + 16°\,6' = 36°\,6'$

log sec	09259	log tan	86285
	92015		60174
	01274		46459
$y_2 =$	1·0297		61539
			29147
		$x_2 =$	−·32392

log = 51044

01274	02046
51044	01274
50230	03320

$\phi_1 = 72^\circ\ 33'$ $r_2 = 1{\cdot}0794$

·0794 log = 89982 ·94003 log = 97314

1·3239 log = 12185 comp = 02686

77797

$A = -\,{\cdot}050975$ $a = 1{\cdot}0638$

1·0638

1·0794

$A - r_1 = -\ {\cdot}0156$

$y_2 = +\ 1{\cdot}0297$

$1{\cdot}0141 = a \cos \chi'$

$-1{\cdot}0453 = a \cos \chi$

00608	01924
02686	02686
97922	99238

$\chi' = 17^\circ\ 35'$ $\chi\ (= \text{Supp. } 10^\circ\ 42') = 169^\circ\ 18'$ $\chi - \chi' = 151^\circ\ 43'$

151°	2·63544		
$43'$	·01250		
$-\sin\chi$		$-\,{\cdot}18566$	
$\sin\chi'$	·30209		02686
			01343
	2·95003		47712
	·18566		
			51741
	2·76437	log = 44160	
		02686	$3a^{\frac{3}{2}} = 3{\cdot}2916$
		01343	1·4482
		67894	
			1·8434
	$T_{12} = 1{\cdot}4482$	·16083	$T_{12} = T_{21} = {\cdot}9217$

82. For the Time in a hyperbola, we have

$$T_{12} = T_{21} = \frac{3}{2\pi}\, a^{\frac{3}{2}} \left\{ e \tan u_2 - h\,.\,l\,.\tan\left(45^\circ + \tfrac{1}{2} u_2\right) \right\},$$

where

$$\tan u_2 = \frac{y_2}{a\sqrt{e^2 - 1}}.$$

Taking as a specimen the case $c = 75^\circ$, we have here

$a =$ ·9004	$e =$ 2·1106	$y_2 =$ 43·341
log = ·95444	log = ·32441	log = 1·63690
$a(e^2 - 1) =$ 3·1106		
log " = ·49284		

and then the calculation is

log a	= 95444		
" $a(e^2-1)$	= 49284		
	44728		
" $a\sqrt{e^2-1}$	= 22364		
log y_2	= 63690		
log tan u	= 41326	$u = 87^\circ\ 47'$	
	32441	$h\,.\,l \tan(45^\circ + \frac{1}{2}u)$	= 3·95140
	73767	$e \tan u$	= 54·660
			3·951
			50·709
		log	= 70508
			95444
			47722
			67894
			31568
		$T_{12} = T_{21}$	= 20·686

83. In the case of the parabola $p = 1$, and the expression for the Time, is

$$T_{12} = T_{21} = \frac{1}{4\pi}\left\{ (\rho + \rho' + \gamma)^{\frac{3}{2}} - (\rho + \rho' - \gamma)^{\frac{3}{2}} \right\},$$

where for

$c = 46^\circ\ 48'$	we have	$T_{12} = T_{21} =$ ·787,
$c = 87^\circ\ 6'$		$T_{12} = T_{21} =$ 2·588.

Planogram No. 1, first part, b = 90°.

	e	x_1	y_1	θ_1	r_1	A	a	T_{11}	T_{12}	T_{21}
Circle.	0°	− ·500	+ ·866	60°	1·000	0	1·000	1·000	1·000	1·000
Ellipses.	5	·461	·892	62° 39	1·004	− ·003	1·003	..		
	10	·420	·927	65 38	1·017	·012	1·012	·960	1·135	·960
	15	·374	·972	68 56	1·041	·030	1·031	..		
	20	·324	1·030	72 33	1·079	·060	1·064	·922	1·448	·922
	25	·267	1·104	76 26	1·136	·107	1·120			
	30	·200	1·200	80 32	1·216	·180	1·220	·887	2·275	·887
	35	·120	1·325	84 49	1·330	·295	1·418			
	40	− ·021	1·491	90 48	1·492	·482	1·931	·838	6·371	·838
	40° 54'	·000	1·515	90	1·515	·515	2·061			
	45	+ ·109	1·722	93 37	1·725	·814	5·362			
Parab.	46° 48'	·166	1·826	95 11	1·834	1·000	∞	·787	∞	·787
Hyperb.	50	·287	2·054	97 57	2·074	1·505	1·981	·750	~	·750
	52° 45'	·418	2·306	100 16	2·344	2·309	·764		~	
	55	·552	2·569	101 8	2·627	− 3·632	·380	·628	~	·628
Line.	60	1·000	3·464	106 6	3·605	− ∞ + ∞	·000	·000	~	·000
Convex.	65	1·917	5·378	109 48	5·716	+ 5·031	·166	Convex orbit.		
	70	+ 5·248	12·233	113 13	13·311	2·898	·257			
	73° 54'	+ ∞ − ∞	∞	115 39 64 21	∞	+ 2·309 − 2·309	·302 ·764	∞	~	∞
Hyperb.	75	− 21·432	43·341	63 42	48·346	2·111	·900	20·68	~	20·68
	80	4·356	7·830	60 55	8·960	1·486	2·056	3·856	~	3·856
	85	2·653	4·322	58 27	5·072	1·115	8·718	2·912	~	2·912
Parab.	87° 6'	2·320	3·644	57 31	4·310	1·000	∞	2·588	∞	2·588
Ellipse.	90	− 2·000	3·000	56 18	3·606	− ·869	7·622	2·255	58·62	2·255

The mark ~ in the T_{12} column shows that there is no Time T_{12}.

Planogram No. 1, second part, b = 270.

	e	x_3	y_3	θ_3	r_3	A	a	T_{12}	T_{23}	T_{13}
Circ.	0°	− ·500	+ ·866	60° ′	1·000	0	1·000	1·000	1·000	1·000
All Ellipses.	5	·537	·848	57 39	1·003	− ·002	1·002			
	10	·573	·837	55 37	1·014	·009	1·009	1·044	·951	1·044
	15	·608	·832	53 52	1·030	·019	1·019			
	20	·643	·834	52 23	1·053	·032	1·033	1·091	·969	1·091
	25	·678	·842	51 10	1·081	·048	1·051			
	30	·714	·857	50 11	1·116	·068	1·073	1·145	1·043	1·145
	35	·752	·879	49 27	1·157	·090	1·098			
	40	·793	·910	48 51	1·207	·115	1·130	1·207	1·192	1·207
	45	·836	·950	48 40	1·266	·145	1·169			
	50	·884	1·002	48 36	1·336	·179	1·217	1·283	1·464	1·283
	55	·938	1·069	48 45	1·442	·218	1·278			
	60	1·000	1·154	49 7	1·527	·264	1·358	1·377	1·983	1·377
	65	1·074	1·266	49 42	1·660	·318	1·466			
	70	1·164	1·412	50 31	1·830	·383	1·622	1·506	3·036	1·506
	75	1·280	1·611	51 35	2·056	·464	1·864			
	80	1·431	1·892	52 53	2·372	·564	2·295	1·771	6·888	1·771
	85	1·651	2·311	54 28	2·840	·694	3·269			
	90	− 2·000	+ 3·000	56 18	3·606	− ·869	7·622	2·255	58·62	2·255

Article Nos. 84 *to* 94. *Planogram No.* 2, *the Meridian* 0°–180° (see Plate III.)

84. The orbit-plane here rotates about an axis in the plane of the ecliptic at right angles to S 1 (Fig. 6). The entire meridian is given by $b = 0°$, $c = 0°$ to $90°$, and $b = 180°$, $c = 0°$ to $90°$, but it is sufficient to consider one of these half meridians, say the latter of them, as the series of values is the same for each of them, with only an interchange of the points 2, 3. I write therefore, $b = 180°$, so that we have

$$\begin{aligned} \alpha\ ,\ \beta\ ,\ \gamma\ &= \quad 0, \quad 1, \quad 0, \\ \alpha',\ \beta',\ \gamma' &= -\cos c,\ 0,\ -\sin c, \\ \alpha'',\ \beta'',\ \gamma'' &= -\sin c,\ 0, \quad \cos c, \end{aligned}$$

consequently

$$x_1' : y_1' : 1 = \qquad \sin c : \quad -\sqrt{3} : \qquad \sqrt{3}\cos c.$$

$$x_2' : y_2' : 1 = -3\cos c - 2\sin c : \quad -\sqrt{3} : \quad -\sqrt{3}\sin c - 2\sqrt{3}\cos c,$$

$$x_3' : y_3' : 1 = \quad 3\cos c - 2\sin c : \quad -\sqrt{3} : \quad \sqrt{3}\sin c - 2\sqrt{3}\cos c$$

and moreover $x' = y$, $y' = -x$; so that, introducing into the formulæ (x_1, y_1), &c., in place of the (x_1', y_1'), &c., we have

$$x_1 = \sec c, \qquad\qquad y_1 = \frac{1}{\sqrt{3}}\tan c,$$

$$x_2 = \frac{1}{\sin c + 2\cos c},\ y_2 = \frac{1}{\sqrt{3}}\,\frac{2\sin c + 3\cos c}{\sin c + 2\cos c},$$

$$x_3 = \frac{-1}{\sin c - 2\cos c},\ y_3 = \frac{1}{\sqrt{3}}\,\frac{2\sin c - 3\cos c}{\sin c - 2\cos c},$$

which, putting

$$\cos\aleph = \frac{2}{\sqrt{5}}, \quad \sin\aleph = \frac{1}{\sqrt{5}}, \quad \tan\aleph = \frac{1}{2}, \quad \aleph = 26°\ 34',$$

become

$$x_1 = \sec c, \qquad\qquad y_1 = \frac{1}{\sqrt{3}}\tan c,$$

$$x_2 = -\frac{1}{\sqrt{5}}\sec(c - \aleph),\ y_2 = \frac{1}{\sqrt{3}}\left\{\frac{8}{5} + \frac{1}{5}\tan(c - \aleph)\right\}.$$

$$x_3 = -\frac{1}{\sqrt{5}}\sec(c+\varpi),\ y_3 = \frac{1}{\sqrt{3}}\left\{-\frac{8}{5}+\frac{1}{5}\tan(c+\varpi)\right\};$$

so that the guide hyperbolas are

$x_1^2 - 3y_1^2 = 1,$	$\frac{1}{2}$ angle of asymptotes $= 30°$			
$x_2^2 = 15y_2^2 - 16\sqrt{3}\,y_2 + 13,$	"	"	$\tan^{-1}\frac{1}{\sqrt{15}} = 14°\ 28'$	
$x_3^2 = 15y_3^2 + 16\sqrt{3}\,y_3 + 13$	"	"	"	"

It is easy to verify that

Hyperbola 2 passes through $x_2 = -\frac{1}{2},\ y_2 = \frac{\sqrt{3}}{2}$, and touches there circle $x^2 + y^2 = 1$,

" 3 " $x_3 = -\frac{1}{2},\ y_3 = -\frac{\sqrt{3}}{2}$ " " "

and we thus have the figure in the Plate.

85. The figure shows the motion of the points 1, 2, 3, along their respective hyperbolas, viz. $c = 0$ to $90°$, the point 1 moves from contact with the circle, along a half branch to infinity: 2 moves from contact along a small portion of the half branch; 3 moves from contact, along the half branch to infinity for $c = \tan^{-1} 2 = 63°\,26'$, and then reappearing at the opposite infinity, as c increases to $90°$ describes a portion of the opposite half branch.

86. For $c = 0$, the orbit is the circle; as c increases the orbit becomes elliptic; then parabolic, $c = 51°$, and afterwards hyperbolic (concave); until for $c = 60°$, the three points are on the horizontal line of the figure, and the orbit is this right line; it is to be noticed that the arrangement of the points on these orbits is 1, 2, 3; so that for the parabola, $T_{,,}$ is $= \infty$, and for the hyperbolas and right line $T_{,,}$ does not exist.

87. For $c < 60°$ until $c = 63°\,26'$ the orbit is a convex hyperbola, the arrangement of the points being still 1, 2, 3: say for $c = 63°\,26' - \epsilon$, the orbit is the convex hyperbola Ω. At $c = 63°\,26'$ there is an abrupt change of orbit; say for $c = 63°\,26' + \epsilon$ the orbit is a concave hyperbola $\Omega_{,}$; and as for $c = 65°\,52'$ the orbit is a parabola; the arrangement of the points on these orbits is 2, 1, 3; so that for the hyperbolas $T_{,,}$ does not exist,

and T_{31} is $= \infty$ for the parabola. Observe also that for the hyperbola Ω_1, the point 3 is at infinity, or we have $T_{1.} = \infty$. As c continues to increase, the orbit becomes an ellipse, the excentricity having a minimum value $=$ ·618 (about), for $c = 69°$ (about). For $c = 89° 25'$ the orbit is again a parabola, and then until $c = 90°$ it is a hyperbola; the order of the points on the last-mentioned parabola and hyperbolas being 1, 3, 2; so that for the parabola $T_{.1}$ is $= \infty$, and for the hyperbolas $T_{.1}$ does not exist. In the hyperbola for $c = 90°$ say the hyperbola Ω', the point 1 is at infinity, or we have $T_{.1} = \infty$. The foregoing results, obtained (except as to the numerical values) by consideration of the figure, will be confirmed by means of the calculated values of c.

88. The equation of the orbit may be written—

$$\begin{vmatrix} \frac{r}{\sqrt{3}} , & \frac{x}{\sqrt{3}}, & y , & \frac{1}{\sqrt{3}} \\ r_1 \cos c , & 1, & \sin c , & \cos c \\ r_2 (\sin c + 2\cos c), & -1, & 2\sin c + 3\cos c, & \sin c + 2\cos c \\ r_3 (\sin c - 2\cos c), & 1, & -2\sin c + 3\cos c, & \sin c - 2\cos c \end{vmatrix} = 0,$$

or developing, this is

$$\begin{aligned} &\frac{r}{\sqrt{3}}\, 6\,(\sin^2 c - 3\cos^2 c), \\ -&\frac{x}{\sqrt{3}}\,\{\quad 4\, r_1 (\sin^2 c - 3\cos^2 c)\cos c \\ &\qquad - \;\; r_2 (\sin c + 2\cos c)(\sin^2 c - 3\cos^2 c) \\ &\qquad + \;\; r_3 (\sin c - 2\cos c)(\sin^2 c - 3\cos^2 c)\} \\ +\,& y \quad \{-2\, r_1 \sin c \cos c \\ &\qquad + r_2 (-\sin^2 c + \sin c\cos c + 6\cos^2 c) \\ &\qquad + r_3 (\sin^2 c + \sin c\cos c - 6\cos^2 c)\} \\ -&\frac{1}{\sqrt{3}}\,\{\quad r_1 . -6\cos^2 c \\ &\qquad + 3\, r_2 (\sin^2 c + \sin c\cos c - 2\cos^2 c) \\ &\qquad + 3\, r_3 (\sin^2 c - \sin c\cos c - 2\cos^2 c)\} = 0; \end{aligned}$$

(observe that the orbit will be a right line if $\sin^2 c - 3\cos^2 c = 0$, that is if $c = 60°$, which is right, since $60°$ is the angular radius of the regulator circle).

89. Putting in the equation $\tan c = \lambda$, and therefore $\cos c = \frac{1}{\sqrt{1+\lambda}}$, the equation becomes

$$r = \frac{1}{6\sqrt{1+\lambda^2}}\Big(4r_1 - (\lambda+2)r_2 + (\lambda-2)r_3\Big)x$$

$$+ \frac{1}{2\sqrt{3}(\lambda^2-3)}\Big(2\lambda r_1 + (\lambda+2)(\lambda-3)r_2 - (\lambda+3)(\lambda-2)r_3\Big)y$$

$$+ \frac{1}{2(\lambda^2-3)}\Big(-2r_1 + (\lambda-1)(\lambda+2)r_2 + (\lambda+1)(\lambda-2)r_3\Big).$$

We have

$$x_1 = \sqrt{1+\lambda^2}, \quad x_2 = \frac{-\sqrt{1+\lambda^2}}{\lambda+2}, \quad x_3 = \frac{\sqrt{1+\lambda^2}}{\lambda-2},$$

$$y_1 = \frac{1}{\sqrt{3}}\lambda, \quad y_2 = \frac{1}{\sqrt{3}}\frac{2\lambda+3}{\lambda+2}, \quad y_3 = -\frac{1}{\sqrt{3}}\frac{2\lambda-3}{\lambda-2},$$

and thence, writing for shortness

$$R_1 = \sqrt{1+\tfrac{4}{3}\lambda^2},$$

$$R_2 = \frac{1}{\sqrt{3}}\sqrt{7\lambda^2+12\lambda+12},$$

$$R_3 = \frac{1}{\sqrt{3}}\sqrt{7\lambda^2-12\lambda+12},$$

we have

$$r_1 = R_1,$$
$$r_2(\lambda+2) = R_2,$$
$$r_3(\lambda-2) = R_3,$$

where r_1, r_2, r_3 are positive, and the signs of R_1, R_2, R_3 must be determined accordingly; viz., R_1 is always positive, and ($c = 0°$ to $c = 90°$, as here supposed) R_2 is also positive; but R_3 has the same sign as $\lambda - 2$; viz., $c = 0°$ to $c = 63° 26'$; R_3 is negative, and $c = 63° 26'$ to $c = 90°$, R_3 is positive. It is to be observed that this position, $c = \tan^{-1} 2 = 63° 26'$, of the pole is the intersection of the meridian $b = 180°$ by a separator circle, and corresponds to an intersection at infinity on the ray 3.

90. Substituting the foregoing values of r_1, r_2, r_3, the equation of the orbit becomes

$$r = \frac{1}{6\sqrt{1+\lambda^2}}(4R_1 - R_4 + R_5)\,x$$

$$+ \frac{1}{2\sqrt{3}(\lambda^2-3)}\{\lambda(2R_1 + R_4 - R_5) - 3(R_4 + R_5)\}\,y$$

$$+ \frac{1}{2(\lambda^2-3)}\{\lambda(R_4 + R_5) - 2R_1 - R_4 + R_5\},$$

where $\lambda = \tan c$; and the equation of the orbit may thence be calculated for any given value of c.

91. The analytical expression for the excentricity is

$$e = \sqrt{A^2 + B^2},$$

where, as above,

$$A = \frac{1}{6\sqrt{1+\lambda^2}}(4R_1 - R_4 + R_5),$$

$$B = \frac{1}{2\sqrt{3}(\lambda^2-3)}\{\lambda(2R_1 + R_4 - R_5) - 3(R_4 + R_5)\};$$

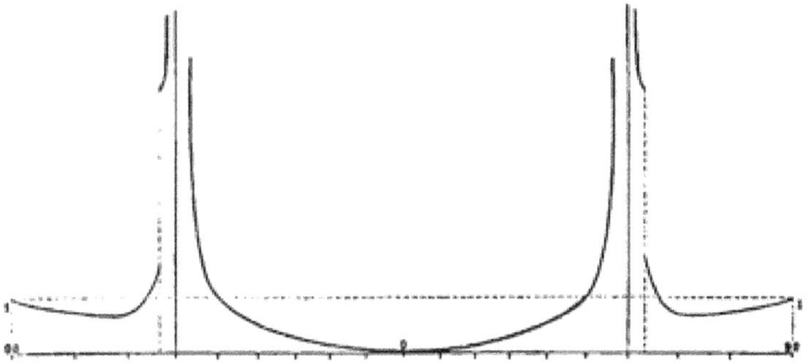

but this expression is too complicated to allow of an analytical discussion of the series of values of e (such as was given for $A_1 = \pm\epsilon$, in planogram No. 1). The numerical calculation gives the results mentioned *ante* No. 87, viz., $c = 0$, $e = 0$; $c = 51^\circ$, $e = 1$, $c = 60^\circ$, $e = \infty$, $c = 63^\circ\,26' - \epsilon$, $e = 4{\cdot}912$, $c = 63^\circ\,26' + \epsilon$, $e = 1{\cdot}853$; $c = 69^\circ$, $e = {\cdot}628$ (min.): $c = 89^\circ\,20'$, (viz. $\lambda = 86{\cdot}176$), $e = 1$; $c = 90^\circ$, $e = 1{\cdot}018$; values which are be exhibited in the diagram in the preceding page.

92. It may be further remarked, in reference to the formula

$$r = \mathrm{A}x + \mathrm{B}y + \mathrm{C},$$

that for $c = 60^\circ$, that is $\lambda = \sqrt{3}$, we have A finite, B and C each infinite, but equal and of opposite signs; viz., the equation becomes $r = {\cdot}2242\,x \pm \infty\,(y-1)$, that is $y = 1$, orbit a right line as above.

The abrupt change at $c = 63^\circ\,26'$, $\lambda = 2$, arises from the change of sign of R_1; viz., $c = 63^\circ\,26' - \epsilon$, $\mathrm{R}_1 = -\frac{4}{\sqrt{3}} = -2{\cdot}309$, but $c = 63^\circ\,26' + \epsilon$, $\mathrm{R}_1 = \frac{4}{\sqrt{3}} = +2{\cdot}309$; the two orbits are

$$c = 63^\circ\,26' - \epsilon, \quad r = {\cdot}234\,x + 4{\cdot}906\,y - 3{\cdot}671, \quad e = 4{\cdot}912, \quad a = {\cdot}159,$$
$$c = 63^\circ\,26' + \epsilon, \quad r = {\cdot}578\,x - 1{\cdot}761\,y + 3{\cdot}257, \quad e = 1{\cdot}853, \quad a = 1{\cdot}338.$$

For $c = 90^\circ$ the equation is

$$r = \frac{4}{3\sqrt{3}}\,x + \frac{2}{3}\,y + \sqrt{\frac{7}{3}}$$
$$= {\cdot}770\,x + {\cdot}666\,y + 1{\cdot}527$$

and therefore $e = \sqrt{\frac{28}{27}} = 1{\cdot}018$ as above; $a = 9\sqrt{21} = 41{\cdot}243$.

It is to be added that for c nearly $= 90^\circ$, or λ very large, we have

$$\mathrm{R}_1 = \frac{2}{\sqrt{3}}\,\lambda, \quad \mathrm{R}_2 = \sqrt{\frac{7}{3}}\,\lambda + 2\sqrt{\frac{3}{7}}, \quad \mathrm{R}_3 = \sqrt{\frac{7}{3}}\,\lambda - 2\sqrt{\frac{3}{7}},$$

and thence

$$\mathrm{A} = \frac{4}{3\sqrt{3}} - \frac{2}{\sqrt{21}}\,\frac{1}{\lambda} = {\cdot}770 - {\cdot}430\,\frac{1}{\lambda}.$$

$$B = \frac{2}{3} - \frac{5}{\sqrt{7}}\,\frac{1}{\lambda} = \cdot 666 - 1\cdot 890\,\frac{1}{\lambda},$$

$$C = \sqrt{\frac{7}{3}} - \frac{1}{\sqrt{3}}\,\frac{2}{\lambda} = 1\cdot 527 - 1\cdot 555\,\frac{1}{\lambda}.$$

It was, in fact, by means of these expressions that the value $\lambda = 86\cdot 176$ ($e = 89°\ 20'$) corresponding to the last-mentioned parabolic orbit was obtained.

93. For the calculation of the table we have

$$\begin{aligned}
\log x_1 &= \overline{10} + \log\sec e,\\
\log y_1 &= \overline{10}\cdot 76144 + \log\tan e,\\
\log x_2 &= \overline{10}\cdot 65052 + \log\sec(e - 26°\ 34'),\\
\log(y_2 - \cdot 92376) &= \overline{10}\cdot 06247 + \log\tan(e - 26°\ 34'),\\
\log x_3 &= \overline{10}\cdot 65052 + \log\sec(e + 26°\ 34'),\\
\log(y_3 + \cdot 92376) &= \overline{10}\cdot 06247 + \log\sec(e + 26°\ 34'),
\end{aligned}$$

the values of r_1, r_2, r_3, are then calculated from

$$x_1 = r_1\cos\phi_1,\quad y_1 = r_1\sin\phi_1,$$

or say

$$\frac{y_1}{x_1} = \tan\phi_1,\qquad r_1 = x_1\sec\phi_1,\ \&c.$$

and those of the chords $\gamma_{12}, \gamma_{23}, \gamma_{31}$, from

$$x_1 - x_2 = \gamma_{12}\cos\theta_{12},\quad y_1 - y_2 = \gamma_{12}\sin\theta_{12},$$

or say

$$\tan\theta_{12} = \frac{x_1 - x_2}{y_1 - y_2},\qquad \gamma_{12} = (x_1 - x_2)\sec\theta_{12}.$$

We have then to find the equation of the orbit $r = Ax + By + C$; this might be done by substituting in the determinant expression the numerical values of $x_1, y_1, r_1, x_2, y_2, r_2, x_3, y_3, r_3$, and so calculating the result, but I have preferred to employ the formula of No. 90, using only the calculated values of r_1, r_2, r_3; viz. we have

$$\begin{aligned}
r_1 &= R_1,\\
r_2(\lambda + 2) &= R_2,\\
r_3(\lambda - 2) &= R_3,
\end{aligned}$$

which gives the values of R_1, R_2, R_3. And then we have e, ϖ, a, from the equations

$$A = e\cos\varpi,\ B = e\sin\varpi,\ a = \frac{\pm C}{1 - e^2}.$$

e and a being each regarded as positive. The times in the elliptic, and parabolic orbits are then calculated from LAMBERT's equation, as explained in regard to Planogram No. 1, but for the hyperbolic orbits, the other formulæ were made use of.

94. I annex a specimen; the characteristics of the logarithms are omitted.

$c = 20°$.

20°		−6° 34′		+46° 34′	
02701	56107	00286	06113	16272	02376
	76144	65052	06247	65052	06247
	32251	65338	12360	81324	08623
$x_1 = 1{\cdot}06418$	$y_1 = {\cdot}21014$	$x_4 = -{\cdot}45017$	·92367	$x_3 = -{\cdot}65049$	·92367
32251	02701		·01329		·12196
02701	00830	$y_4 = +$	·91038	$y_3 =$	·80171
29950	03531	log =	·95922	$\log y_3 =$	90402
$\phi_1 = 11° 10′$	$r_1 = 1{\cdot}0847$	95922	95922	90402	90402
		65338	04752	81324	10980
		30584	00674	09078	01382

$\phi_4 (= 63° 41′) = 116° 19′,\ r_4 = 1{\cdot}0157 \quad \phi_3 (= 50° 57′) = 230° 57′,\ r_3 = 1{\cdot}0323$

The calculation of the equation of the orbit is then as follows—

$\lambda = \cdot 36397$
$\log = 56107$
12214
$\lambda^2 = \cdot 13248$
$\lambda^2 - 3 = -2\cdot 86752$
$\log = 45750$

$\log \sqrt{1 + \lambda^2} = 02701$
77815

$\log 6 \sqrt{1 + \lambda^2} = 80516 \ (a)$

45750
30103

$\log 2 (\lambda^2 - 3) = 75853 \ (c)$
23856

$\log 2 \sqrt{3} (\lambda^2 - 3) = 99709 \ (b)$

$\log R_1 = 03531$
$R_1 = 1\cdot 0847$
$\lambda + 2 = 2\cdot 36397$
$\log = 37364$
$\log r_2 = 00674$

38038

$R_2 = + 2\cdot 4010$

$\lambda - 2 = - 1\cdot 63603$
$\log = 21378$
$\log r_3 = 01382$

22760

$R_3 = - 1\cdot 6889$

$4 R_1 = 4\cdot 3388$
$- R_2 \qquad - 2\cdot 4010$
$+ R_3 \qquad - 1\cdot 6889$

$4\cdot 0899$

$\cdot 2489$

$\log = 39602$
$(a) = 80516$

59086

$A = \cdot 038982$

$2 R_1 = 2\cdot 1694$
$+ R_2 \quad 2\cdot 4010$
$- R_3 \quad 1\cdot 6889$

$6\cdot 2593^*$

$\log = 79653$
$\lambda = 56107$

35760

$+ 2\cdot 2782$
$- 3 R_2 \qquad - 7\cdot 2032$
$- 3 R_3 \quad 5\cdot 0667$

$7\cdot 3449$
$- 7\cdot 2032$

$0\cdot 1417$

$\log = 15137$
$(b) = 99709$

$\cdot 15428$

$B = - \cdot 014265$

$R_2 = 2\cdot 4010$
$R_3 = - 1\cdot 6889$

$+ \cdot 7121$

$\log = 85254$
$\lambda = 56107$

41361

$+ \cdot 25919$
$- 6\cdot 2593^*$

$- 6\cdot 00011$

$\log = 77818$
$(c) = 75853$

$\cdot 01965$

$C = + 1\cdot 0464$

$$\log B = 15428 \qquad 02729$$
$$\log A = \underline{59086} \qquad \underline{59086}$$
$$56342 \qquad 61815$$
$$\varpi\ (= 20^\circ\ 6') = 160^\circ\ 52' \qquad e = \cdot 04151$$
$$13630$$
$$e^2 = \cdot 001723 \qquad \log C = 01965$$
$$1 - e^2 = \cdot 998277 \qquad \log = \underline{99925}$$
$$a = 1\cdot 0481 \qquad 02040$$

The calculation of the Times is similar to that for the first planogram, and requires no further illustration.

The Table for Planogram No. 2 is as follows —

Planogram No. 2,

	ϵ	x_1 all +	y_1 all +	x_2 all −	y_2 all +	x_3	y_3	r_1	θ_1	r_2	θ_2	r_3	θ_3
Circle	0	1·000	·000	·500	·866	− ·500	− ·866	1·000	0° 0′	1·000	120° 0′	1·000	240° 0′
	5	1·004	·051	·481	·878	·525	·853	1·005	2 52	1·001	118 42	1·001	238 23
	10	1·015	·102	·467	·889	·557	·838	1·020	5 44	1·004	117 41	1·006	236 34
	15	1·035	·155	·456	·900	·598	·821	1·047	8 30	1·009	116 54	1·016	233 58
	20	1·064	·210	·450	·910	·650	·802	1·085	11 10	1·015	116 19	1·032	230 57
Ellipses	25	1·103	·269	·447	·921	·719	·778	1·136	13 43	1·024	115 55	1·060	227 15
	30	1·155	·333	·448	·931	·812	·749	1·202	16 6	1·033	115 42	1·104	222 42
	35	1·221	·404	·452	·941	·939	·710	1·286	18 19	1·044	115 40	1·178	217 6
	40	1·305	·484	·460	·951	1·125	·657	1·392	20 21	1·056	115 48	1·302	210 19
	45	1·414	·577	·471	·962	1·414	·577	1·527	22 12	1·071	116 6	1·528	202 12
	50	1·556	·688	·487	·974	1·925	·440	1·701	23 51	1·088	116 35	1·975	192 53
Parab.	51° 0′	1·589	·713	·491	·976	2·077	·400	1·741	24 9	1·093	116 43	2·115	190 54
	52	1·624	·739	·495	·978	2·256	·353	1·787	24 28	1·097	116 51	2·283	188 53
	54	1·701	·795	·504	·984	2·729	·229	1·878	25 2	1·105	117 7	2·738	184 48
Hyperbs.	55	1·743	·824	·509	·986	3·049	− ·145	1·928	25 18	1·109	117 26	3·053	182 43
	56 18	1·802	·866	·515	·990	3·601	·000	1·999	25 39	1·116	117 30	3·601	180 0
	59	1·942	·961	·530	·997	5·786	+ ·566	2·166	26 20	1·129	118 59	5·813	174 25
Line	60	2·000	1·000	·536	1·000	7·468	1·000	2·236	26 34	1·134	118 11	7·534	172 23
Convex	61	2·063	1·042	·542	1·003	− 10·53	+ 1·793	2·311	26 48	1·140	118 24	10·68	170 20
	63° 26′ − ε 63° 26′ + ε	2·236	1·155	·559	1·010	− ∞ + ∞	+ ∞ − ∞	2·517	27 19	1·155	118 57	∞	165 31 345 31
Hyperbs.	64	2·281	1·184	·563	1·012	+ 45·22	− 12·60	2·570	27 26	1·157	119 6	46·94	344 26
	65	2·366	1·238	·571	1·015	16·36	5·146	2·670	27 37	1·165	119 21	17·15	342 37
Parab.	65 52	2·446	1·289	·578	1·019	10·12	3·552	2·765	27 47	1·171	119 35	10·80	340 56
	66	2·459	1·297	·579	1·019	9·987	3·500	2·779	27 48	1·172	119 37	10·59	340 41
	68	2·669	1·429	·596	1·026	5·617	2·369	3·028	28 10	1·186	120 21	6·090	337 8
	70	2·924	1·586	·616	1·033	3·912	1·927	3·316	28 29	1·202	120 48	4·360	333 46
Ellipses	72	3·237	1·777	·638	1·041	3·008	1·694	3·693	28 47	1·221	121 29	3·455	330 38
	75	3·864	2·155	·674	1·054	2·230	1·488	4·424	29 9	1·251	122 36	2·682	326 17
	80	5·759	3·274	·751	1·079	1·568	1·312	6·624	29 37	1·315	124 49	2·045	320 5
	85	11·47	6·599	·854	1·112	1·217	1·216	13·25	29 54	1·402	127 32	1·720	315 1
Parab.	89° 20′	86·41	49·79	·979	1·148	1·024	1·161	99·5	29 56	1·508	130 24	1·548	311 24
Hyperbs.	90 − ε	∞	∞	1·000	1·155	+ 1·000	− 1·155	∞	30 0	1·527	130 54	1·527	310 54

$b = 180^\circ,\ c = 0^\circ$ to 90°.

Equation of Orbit.			e	ϖ	a	γ_{12}	γ_{23}	γ_{13}	T_{12}	T_{23}	T_{13}	
$r = Ax$	$+ By$	$+ C$		° ′								
·000	·000	+ 1·000	·000	ind.	1·000	1·732	1·732	1·732	1·000	1·000	1·000	0°
												5
+·0101	− ·0015	1·0104	·010	171 19	1·010	1·729	1·832	1·678	·987	1·106	·953	10
												15
·039	·014	1·046	·041	160 52	1·048	1·724	1·991	1·668	·956	1·316	·946	20
												25
·083	·061	1·126	·103	143 48	1·138	1·718	2·244	1·710	·924	1·777	·962	30
												35
·135	·209	1·317	·248	122 54	1·404	1·740	2·684	1·826	·878	3·238	·966	40
·161	·395	1·527	·426	112 12	1·867	1·805	3·055	1·925				45
·186	·815	1·972	·836	102 50	6·554	2·016	3·659	2·063	·878	48·60	·849	50
·191	·982	2·140	1·000	101 14	∞	2·100	3·831	2·097	·879	∞	·820	51° 0′
·196	1·150	2·319	1·166	99 39	6·434							52
·203	1·719	2·898	1·720	96 44	1·481							54
·207	2·182	3·366	2·192	95 25	·885	2·781	4·890	2·258	·895	~	·665	55
·212	3·074	4·227	3·081	93 56	·498							56° 18′
·221	− 14·15	+ 15·42	14·15	90 30	·077							59
·224	± ∞ (y − 1)		∞	90 0	·000	6·932	9·468	2·536	·000	~	·000	60
									Convex Orbits.			61
·234	+ 4·906	− 3·671	4·912	87 17	·159							63° 26′ − 1
·578	− 1·761	+ 3·257	1·853	108 11	1·338	∞	∞	2·799	~	∞	·909	63° 26′ + 1
												64
·587	·979	2·494	1·134	120 21	8·666	18·014	15·380	2·945				65
·591	·805	2·257	1·000	126 19	∞	11·633	9·072	3·036	∞	7·746	1·386	65° 52′
·593	·779	2·221	·979	127 15	53·83							66
·606	·338	1·894	·693	150 53	3·645							68
·619	− ·120	1·708	·630	169 0	2·834	5·409	3·649	3·584	5·735	6·343	2·685	70
·635	+ ·027	1·599	·636	182 25	2·674							72
·654	·185	1·497	·680	195 47	2·783	3·859	3·981	4·644	2·62	6·68	4·64	75
·692	·366	1·439	·783	207 52	3·716	3·327	6·212	6·870	1·97	10·21	9·343	80
·740	·514	1·455	·892	214 47	7·721	3·115	12·895	13·480				85
·764	·645	1·505	1·000	219 51	∞	3·055	99·43	102·7	1·200	225·4	∞	89° 20′
+ ·770	+ ·666	+ 1·527	1·018	220 6	41·000	3·055	∞	∞	1·148	∞	~	90 − 1

The mark ~ in any of the T columns shows that the Time does not exist.

Article Nos. 95 to 98. Planogram No. 3, the Orbit-pole at one of the points A.

95. When the orbit-pole is at one of the points A, the orbit-plane passes through one of the rays, and as there is no longer on this ray any determinate point of intersection, the orbit (as was seen) becomes indeterminate. Thus consider the point A for which $b = 270°$, $c = 60°$: we have

$$\alpha,\ \beta,\ \gamma = -1, \quad 0, \quad 0,$$

$$\alpha',\ \beta',\ \gamma' = 0, \quad -\tfrac{1}{2}, \quad -\frac{\sqrt{3}}{2},$$

$$\alpha'',\ \beta'',\ \gamma'' = 0, \quad -\frac{\sqrt{3}}{2}, \quad \frac{1}{2},$$

and consequently the formula gives

$$x_1' : y_1' : 1 = 0 : 0 : 0,$$

$$x_2' : y_2' : 1 = -\tfrac{1}{2}\sqrt{3} - \sqrt{3} : 3 : -\tfrac{1}{2}\sqrt{3} - \sqrt{3},$$

$$x_3' : y_3' : 1 = -\tfrac{1}{2}\sqrt{3} - \sqrt{3} : -3 : -\tfrac{1}{2}\sqrt{3} - \sqrt{3},$$

and, moreover, $x = -x'$, $y = -y'$. From the formula the value of x_1' or x_1 is given as $\frac{0}{0}$, but the true value is obviously $x_1 = 1$; the value of y_1 is actually indeterminate. The formulæ give the values of (x_2, y_2), (x_3, y_3), viz. the system is

$$x_1 = 1, \qquad y_1 = \text{ind.}$$

$$x_2 = -1, \qquad y_2 = \frac{2}{\sqrt{3}}, \quad \text{whence } r_2 = r_3 = \sqrt{\frac{7}{3}},$$

$$x_3 = -1, \qquad y_3 = -\frac{2}{\sqrt{3}},$$

so that the orbits in the planogram are the whole series of conics having a given focus, S, and passing through two fixed points, 2, 3, having the common abscissa $x = -1$, and at equal distances $\frac{2}{\sqrt{3}}$ $(= 1 \cdot 15470)$ on opposite sides of the axis. The axis of x is obviously the common transverse axis for all the orbits; that is, the equation of the orbit will be of the form $r = Ax + B$; and writing $x = -1$, we have $\sqrt{\frac{7}{3}} = -A + B$, viz. the equation is $r - \sqrt{\frac{7}{3}} = A\,(x+1)$; the value of A will be determined if we assume for the point 1 a determinate position on the line $x = 1$, say its ordinate is $= y_1$; for then if $r_1 = \sqrt{1 + y_1^2}$ we have $r_1 - \sqrt{\frac{7}{3}} = 2A$, and the equation is $r - \sqrt{\frac{7}{3}} = \frac{1}{2}(r_1 - \sqrt{\frac{7}{3}}\,(x+1)$. In particular if $y_1 = 0$, we have $r_1 = 1$, and the equation of the orbit is $r - \sqrt{\frac{7}{3}} = \frac{1}{2}\left(1 - \sqrt{\frac{7}{3}}\right)(x+1)$: this is the orbit, excentricity $\frac{1}{2}\left(\sqrt{\frac{7}{3}} - 1\right)$, $= \cdot 264$, belonging to the point A as a point in planogram No. 1: for the value of y_1 being in that planogram originally assumed $= 0$, is of course $= 0$ when the orbit-pole comes to be the point A.

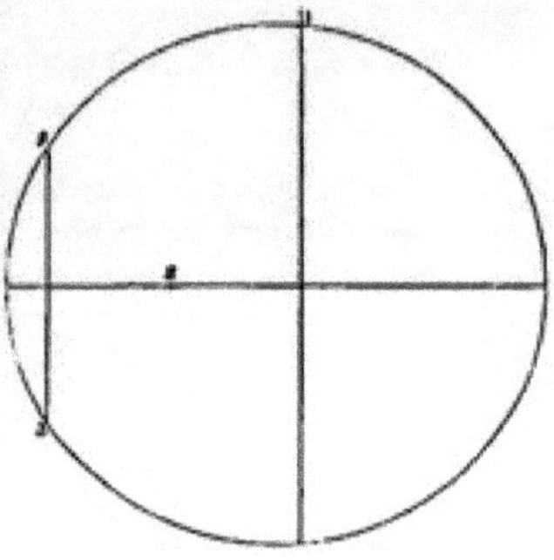

Fig. 9.

96. We may conversely take the equation of the orbit, or say the value of A $(= \pm e)$ in the equation $r - \sqrt{\frac{7}{3}} = A\,(x+1)$, to be given; and then writing $x = x_1 = 1$, we have

$$r_1 = \sqrt{\frac{7}{3}} + 2A, \quad \text{that is} \quad y_1^2 = \left(\sqrt{\frac{7}{3}} + 2A\right)^2 - 1.$$

For

$$r_1 = 1 \text{ or } y_1 = 0, \quad A = \frac{1}{2}\left(1 - \sqrt{\frac{7}{3}}\right) = -\cdot 264.$$

and as r_1 increases to $r_1 = \sqrt{\frac{7}{3}}$, or y_1 increases to $\pm\frac{2}{\sqrt{3}}$, A diminishes from $-\cdot 264$ to 0; viz., for $r_1 = \sqrt{\frac{7}{3}}$, or $y_1 = \pm\frac{2}{\sqrt{3}}$, the orbit is a circle; as r_1 increases from $\sqrt{\frac{7}{3}}$, or y_1 from $\pm\frac{2}{\sqrt{3}}$, A increases from 0 positively; for $r_1 = \sqrt{\frac{7}{3}} + 2$, $= 3\cdot 527$, or $y_1 = \pm\sqrt{\frac{16+4\sqrt{21}}{3}}$, $= \pm 2\cdot 896$, A becomes $= 1$; that is, the orbit is a parabola; and for larger positive values of r_1, or positive or negative values of y_1, the orbit is a hyperbola (concave); and ultimately for $r_1 = \infty$ or $y_1 = \pm\infty$, the orbit is the right line $x + 1 = 0$. Thus A extends from $-\cdot 264$ to 0, and thence from 0 positively to $+\infty$.

97. In further illustration, suppose that the orbit-pole, instead of being at A, is a point in the immediate neighbourhood of A, say that the rectangular spherical co-ordinates, measured from A in the direction of the the meridian and perpendicular thereto, are ξ and η; the colatitude and longitude of the orbit-pole being thus $c = 60° + \xi$, and $b = 270° + \frac{2}{\sqrt{3}}\eta$; we have then, ξ, η being indefinitely small,

$$\alpha,\ \beta,\ \gamma = -1,\quad -\frac{2}{\sqrt{3}}\eta,\quad 0,$$

$$\alpha',\ \beta',\ \gamma' = \frac{1}{\sqrt{3}}\eta,\quad -\frac{1}{2}+\frac{\sqrt{3}}{2}\xi,\quad \frac{\sqrt{3}}{2}-\frac{1}{2}\xi,$$

$$\alpha'',\ \beta'',\ \gamma'' = \eta,\quad -\frac{\sqrt{3}}{2}-\frac{1}{2}\xi,\quad \frac{1}{2}-\frac{\sqrt{3}}{2}\eta;$$

and thence

$$x'_1 = \left(-\frac{1}{2}+\frac{\sqrt{3}}{2}\xi\right)\sqrt{3}+\frac{\sqrt{3}}{2}+\frac{1}{2}\xi = 2\xi$$

$$: y'_1 = : -\frac{2}{\sqrt{3}}\eta\cdot\sqrt{3} \qquad : -2\eta$$

$$: 1 \qquad : -\frac{\sqrt{3}}{2}-\frac{1}{2}\xi+\left(\frac{1}{2}-\frac{\sqrt{3}}{2}\right)\xi \qquad : -2\xi;$$

that is, $x'_1 = -1$, $y'_1 = \frac{\eta}{\xi}$, or what is the same thing, $x_1 = -1$, $y_1 = \frac{\eta}{\xi}$;

the values of x_2, y_2, and x_1, y_1, differ from their former values only by terms in ξ, η, which may be neglected; that is, we have as before $x_2 = -1$, $y_2 = \frac{1}{\sqrt{3}}$ and $x_1 = -1$, $y_1 = -\frac{2}{\sqrt{3}}$; and we thus see that the foregoing determination of the orbit for an arbitrary value of y_2, writing therein $y_2 = -\frac{\eta}{\xi}$ (or what would be the same thing $y_2 = \frac{\eta}{\xi}$) gives the orbit for the neighbouring position $c = 60° + \xi$, and $b = 270° + \frac{2}{\sqrt{3}}\eta$ of the orbit-pole. Writing for greater convenience $\xi = \rho \cos \psi$, $\eta = \rho \sin \psi$, the indefinitely small quantity ρ will denote the distance of the orbit-pole from A, and its azimuth measured from the meridian will be $= \psi$. We then have $y_2 = -\tan \psi$, and $r_2 = \sqrt{1 + y_2^2}, = \pm \sec \psi$, or, if to fix the ideas, ψ be considered as $< \pm 90°$, then $r_2 = \sec \psi$: we have thus ($A = \pm e$ as before) $A = \frac{1}{2}\left(-\sqrt{\frac{7}{3}} + \sec \psi\right)$; viz., observing that $\sqrt{\frac{7}{3}} = 1{\cdot}527$, we obtain

$$\psi = 0, \qquad A = -\tfrac{1}{2}\left(\sqrt{\tfrac{7}{3}} - 1\right) = -{\cdot}264$$

$$\psi = \sec^{-1} \sqrt{\tfrac{7}{3}} = \pm 49° 6', \quad A = 0$$

$$\psi = \sec^{-1}\left(2\sqrt{\tfrac{7}{3}} - 1\right) = \pm 60° 51', \quad A = \tfrac{1}{2}\left(\sqrt{\tfrac{7}{3}} - 1\right) = +{\cdot}264$$

$$\psi = \sec^{-1}\left(\sqrt{\tfrac{7}{3}} + 1\right) = \pm 73° 31', \quad A = 1$$

$$\psi = \pm (90° - i), \quad A = + \infty.$$

98. These results will have to be further considered in reference to the course of the isoeccentric curves through the point A. I remark here that, although it appears that although for excentricities less than ${\cdot}264$, and in particular for the excentricity $= 0$, there are real directions of passage from A to a neighbouring point, yet there are not through A any real branches of the corresponding isoeccentric curves; viz., A is in regard to these curves, an isolated point with *real* tangents; that is a point in the nature of an evanescent lemniscate. As regards the excentricity $= 0$, it is obvious that this must be so; viz., that there can be no real branch through A. In fact, the orbit can only be a circle when the intersection

by the orbit-plane of the hyperboloid which contains the three rays is also a circle; that is, the orbit is a circle only when the orbit-plane coincides with the plane of the ecliptic.

Article Nos. 99 to 103. Planogram No. 4, the Orbit-pole in the Ecliptic.

99. When the orbit-pole describes the circle of the ecliptic, the orbit-plane passes through the axis of z, or polar axis. We have $c = 90°$, and consequently

$$\begin{array}{llll} \alpha,\ \beta,\ \gamma & = \sin b, & -\cos b, & 0, \\ \alpha',\ \beta',\ \gamma' & = 0\ , & 0\ , & -1, \\ \alpha'',\ \beta'',\ \gamma'' & = \cos b, & \sin b, & 0. \end{array}$$

Reverting for a moment to the general case where the six co-ordinates of the ray are (a, b, c, f, g, h), the formulæ for the intersection by the orbit-plane are

$$\begin{array}{llll} x' : y' : 1 = & (a, b, c)(\alpha', \beta', \gamma') & = -c \\ & : -(a, b, c)(\alpha, \beta, \gamma) & : -a \sin b + b \cos b \\ & : \ \ (f, g, h)(\alpha'', \beta'', \gamma'') & : \ \ f \cos b + g \sin b, \end{array}$$

that is

$$\frac{1}{x'} + \frac{f}{c}\cos b + \frac{g}{c}\sin b = 0,$$

$$\frac{y'}{x'} + \frac{b}{c}\cos b - \frac{a}{c}\sin b = 0;$$

and thence

$$1 : \cos b : \sin b = \frac{-af - bg}{c^2} : \frac{gy' + a}{cx'} : \frac{b - fy'}{cx'}$$

$$= \quad hx' \quad : gy' + a : -fy' + b;$$

consequently

$$h^2 x'^2 = (gy' + a)^2 + (fy' - b)^2,$$

or, what is the same thing,

$$h^2 x'^2 = (f^2 + g^2) y'^2 + 2(ag - bf) y' + + a^2 + b^2.$$

or, in particular, if (as in the special symmetrical case) $a g - b f = 0$, then

$$h x'^2 = (f^2 + g^2) y'^2 + a^2 + b^2.$$

100. For the symmetrical system of rays we have as before

$$\begin{array}{llllllll} a_1, b_1, c_1, f_1, g_1, h_1 = & 0, & \sqrt{3}, & -1, & 0, & 1, & \sqrt{3}, \\ a_2, b_2, c_2, f_2, g_2, h_2 = & 3, & \sqrt{3}, & 2, & \sqrt{3}, & 1, & -2\sqrt{3}, \\ a_3, b_3, c_3, f_3, g_3, h_3 = & -3, & \sqrt{3}, & 2, & \sqrt{3}, & 1, & -2\sqrt{3}, \end{array}$$

and thence

$$\begin{array}{l} x'_1 : y'_1 : 1 = \quad 1 : \quad \sqrt{3} \cos \delta \quad : \sin \delta, \\ x'_2 : y'_2 : 1 = -2 : -3 \sin \delta + \sqrt{3} \cos \delta : \sin \delta + \sqrt{3} \cos \delta, \\ x'_3 : y'_3 : 1 = -2 : \quad 3 \sin \delta + \sqrt{3} \cos \delta : \sin \delta - \sqrt{3} \cos \delta, \end{array}$$

or, what is the same thing,

$$\begin{array}{ll} x'_1 = \operatorname{cosec} \delta \quad , & y'_1 = \sqrt{3} \cot \delta, \\ x'_2 = \dfrac{-1}{\sin \delta + \sqrt{3} \cos \delta}, & y'_2 = \dfrac{\sqrt{3} (\cos \delta - \sqrt{3} \sin \delta)}{\sin \delta + \sqrt{3} \cos \delta}, \\ x'_3 = \dfrac{-2}{\sin \delta - \sqrt{3} \cos \delta}, & y'_3 = \dfrac{\sqrt{3} (\cos \delta + \sqrt{3} \sin \delta)}{\sin \delta - \sqrt{3} \cos \delta}; \end{array}$$

or as these may also be written

$$\begin{array}{ll} x'_1 = \operatorname{cosec} \delta \quad , & y'_1 = \sqrt{3} \cot \delta, \\ x'_2 = -\operatorname{cosec} (\delta + 60^\circ), & y'_2 = \sqrt{3} \cot (\delta + 60^\circ), \\ x'_3 = -\operatorname{cosec} (\delta - 60^\circ), & y'_3 = \sqrt{3} \cot (\delta - 60^\circ), \end{array}$$

so that for each of these sets we have

$$x'^2 - \tfrac{1}{3} y'^2 = 1.$$

(The curve is in fact a section of the hyperboloid of revolution, $x^2 + y^2 - \frac{1}{3} z^2 = 1$, which passes through the three rays.)

102. As regards the equation of the orbit I will first consider the particular cases $b = 90^\circ$, $b = 0^\circ$, which should agree with the orbits for $c = 90^\circ$ in the planogramms 1 and 2 respectively.

For $b = 90^\circ$ we have $x' = x$, $y' = y$ and

$$\begin{aligned} x'_1 &= 1, & y'_1 &= 0, \\ x'_2 &= -2, & y'_2 &= -3, \\ x'_3 &= -2, & y'_3 &= 3, \end{aligned}$$

and the orbit is at once found to be

$$r = \frac{1 - \sqrt{13}}{3}(x' - 1),$$

the excentricity (regarded as positive) being thus $\frac{\sqrt{13} - 1}{3}$, $= \cdot 7685$ as before. For $b = 0^\circ$ there is a discontinuity, and I write successively $b = +\epsilon$, and $b = -\epsilon$. For $b = +\epsilon$ we have $x' = -y$, $y' = x$, and

$$\begin{aligned} x'_1 &= \infty, & y'_1 &= \infty\sqrt{3}, \\ x'_2 &= -\frac{2}{\sqrt{3}}, & y'_2 &= -1, \\ x'_3 &= \frac{2}{\sqrt{3}}, & y'_3 &= -1, \end{aligned}$$

and the orbit is found to be

$$r = \frac{2}{3}x' + \frac{4}{3\sqrt{3}}y' + \frac{\sqrt{7}}{\sqrt{3}} = \cdot 666\,x + \cdot 770\,y + 1\cdot 527;$$

and similarly for $b = -\epsilon$ the equation is

$$r = -\frac{2}{3}x' + \frac{4}{\sqrt{3}}y' + \frac{\sqrt{7}}{\sqrt{3}} = -\cdot 666\,x + \cdot 770\,y + 1\cdot 527;$$

hence the excentricity is

$$e = \sqrt{\frac{28}{27}}, \quad = 1\cdot 018, \text{ as before.}$$

102. Considering now the general case where b has any value whatever, the equation of the orbit is

$$\begin{vmatrix} r & , & x', & y' & , & 1 \\ r_1 \sin b & , & 1, & \sqrt{3}\cos b & , & \sin b \\ r_2(\sin b + \sqrt{3}\cos b), & & -2, & -3\sin b + \sqrt{3}\cos b, & & \sin b + \sqrt{3}\cos b \\ r_3(\sin b - \sqrt{3}\cos b), & & -2, & 3\sin b + \sqrt{3}\cos b, & & \sin b - \sqrt{3}\cos b \end{vmatrix} = 0$$

($x' = x \sin b - y \cos b$, $y' = x \cos b + y \sin b$, as before).

The coefficient of r is readily found to be $-6\sqrt{3}(\sin^2 b + \cos^2 b)$, $= -6\sqrt{3}$; hence completing the development, dividing by $6\sqrt{3}$, and transposing, the equation of the orbit is

$$\begin{aligned} r = \ & \frac{1}{6}\Big[2 r_1 \sin b \quad - r_2(\sin b + \sqrt{3}\cos b) - r_3(\sin b - \sqrt{3}\cos b)\Big]\, x' \\ & + \frac{1}{6\sqrt{3}}\Big[4 r_1 \sin b \cos b \quad + r_2\Big(-2\sin b\cos b + \sqrt{3}(\cos^2 b - \sin^2 b)\Big) \\ & \qquad\qquad + r_3\Big(-2\sin b\cos b - \sqrt{3}(\cos^2 b - \sin^2 b)\Big)\Big]\, y' \\ & + \frac{1}{6}\Big[4 r_1 \sin^2 b \quad + r_2\Big(\sin^2 b + 3\cos^2 b + 2\sqrt{3}\sin b\cos b\Big) \\ & \qquad\qquad + r_3\Big(\sin^2 b + 3\cos^2 b - 2\sqrt{3}\sin b\cos b\Big)\Big], \end{aligned}$$

where

$$r_1 = \frac{\sqrt{\sin^2 b + 4\cos^2 b}}{\sin b},$$

$$r_2 = \frac{\sqrt{13\sin^2 b + 7\cos^2 b - 6\sqrt{3}\sin b\cos b}}{\sin b + \sqrt{3}\cos b},$$

$$r_3 = \frac{\sqrt{13\sin^2 b + 7\cos^2 b + 6\sqrt{3}\sin b\cos b}}{\sin b - \sqrt{3}\cos b};$$

in which expressions the signs of the radicals must be such that r_1, r_2, r_3 shall be positive. Hence writing $\tan b = \eta$, $\left(\frac{1}{\cos b} = \sqrt{1 + \eta^2}\right.$, which determines the sign of $\left.\sqrt{1 + \eta^2}\right)$, also

$$R_1 = \sqrt{\eta^2 + 4},\ R_2 = \sqrt{13\eta^2 - 6\sqrt{3}\eta + 7},\ R_3 = \sqrt{13\eta^2 + 6\sqrt{3}\eta + 7}$$

and therefore

$$\eta r_1 = R_1, \quad (\eta + \sqrt{3})\, r_2 = R_2, \quad (\eta - \sqrt{3})\, r_3 = R_3.$$

which last equations determine the signs of R_1, R_2, R_3 respectively, the equation of the orbit is

$$\begin{aligned} r = {} & \frac{1}{6\sqrt{1+\eta^2}} \quad (2R_1 - R_2 - R_3)\,x' \\ & + \frac{1}{6(1+\eta^2)\sqrt{3}} \left(+R_1 + R_2(1-\eta\sqrt{3}) + R_3(1+\eta\sqrt{3})\right) y' \\ & + \frac{1}{6(1+\eta^2)} \quad \left(+R_1\eta + R_2(\eta+\sqrt{3}) + R_3(\eta-\sqrt{3})\right). \end{aligned}$$

Thus if $b = +\epsilon$, then also $\eta = +\epsilon$,

$$\sqrt{1+\eta^2} = 1,\ R_1 = 2,\ R_2 = \sqrt{7},\ R_3 = -\sqrt{7},$$

and the equation is

$$r = \frac{1}{6}\,4x' + \frac{1}{6\sqrt{3}}\,8y' + \frac{1}{6}\,2\sqrt{7}\sqrt{3} = \frac{2}{3}x' + \frac{4}{3\sqrt{3}}y' + \frac{\sqrt{7}}{\sqrt{3}} = \cdot 666\,x' + \cdot 770\,y' + 1\cdot 527$$

as before; and similarly if $b = 90^\circ$.

And moreover, if $b = 30^\circ$, then

$$\eta = \frac{1}{\sqrt{3}},\ R_1 = \sqrt{\frac{13}{3}},\ R_2 = \frac{4}{\sqrt{3}},\ R_3 = -\frac{2\sqrt{13}}{3},$$

whence the equation of the orbit is

$$\begin{aligned} r &= \tfrac{1}{3}(\sqrt{13}-1)\,x' + 0\,y' + \tfrac{1}{3}(\sqrt{13}+2) \\ &= \quad \cdot 868\,x' + 0\,y' + \ 1\cdot 868. \end{aligned}$$

103. The equation of the orbit should be tabulated from $b = 0$ to $b = 30^\circ$, the equations for the remainder of the circumference will be then found by successive repetition of this interval in direct and reverse order, with however a change of sign, in the manner about to be explained,

$$\begin{array}{ll} b = \epsilon, & r = + \cdot 666\,x' + \cdot 770\,y' + 1\cdot 527, \\ b = 30^\circ, & r = + \cdot 868\,x' + \ 0\,y' + 1\cdot 868, \\ b = 60^\circ - \epsilon, & r = + \cdot 666\,x' - \cdot 770\,y' + 1\cdot 527, \\[4pt] b = 60^\circ + \epsilon, & r = - \cdot 666\,x' + \cdot 770\,y' + 1\cdot 527, \\ b = 90^\circ, & r = - \cdot 868\,x' + \ 0\,y' + 1\cdot 868, \\ b = 120^\circ - \epsilon, & r = - \cdot 666\,x' - \cdot 770\,y' + 1\cdot 527, \end{array}$$

$30° + \beta$ same as $30° - \beta$, reversing sign of the y' coefficient.

$90° + \beta$ same as $90° - \beta$, reversing sign of the y' coefficient, and whole interval 60° to 120° same as interval 0° to 60°, except that the signs of the x' coefficient are reversed, and the remaining two intervals, 120° to 240° and 240° to 360°, are merely repetitions of the interval 0° to 120°.

As regards the interval 0° to 30° the only intermediate value that I have calculated is $b = 15°$, viz., we then have

$$b = 15°, \quad r = \cdot 811\, x' + \cdot 403\, y' + 1 \cdot 787.$$

Calculating for the foregoing values $b = 0°$, $b = 15°$, $b = 30°$, the values of e, ϖ, a, these are found to be

$b = 0°$,	$e = 1 \cdot 018$	$\varpi = 120°\ 6'$	$a = 41 \cdot 24$
$b = 15°$,	$e = \cdot 906$	$\varpi = 106°\ 17'$	$a = 10 \cdot 008$
$b = 30°$,	$e = \cdot 868$	$\varpi = 180°$	$a = 7 \cdot 604$

Article Nos. 104 *to* 113. *Planogram No.* 5. *The Orbit-pole on a Separator.*

104. If the orbit-plane rotate round a line parallel to one of the rays, the orbit-pole will describe a separator circle, and conversely. I consider the general case of a ray the six co-ordinates of which are (a, b, c, f, g, h), and for which the intersections with the orbit-plane are given by

$$x' : y' : 1 = (\mathrm{a}, \mathrm{b}, \mathrm{c})(\alpha', \beta', \gamma') : -(\mathrm{a}, \mathrm{b}, \mathrm{c})(\alpha, \beta, \gamma) : (\mathrm{f}, \mathrm{g}, \mathrm{h}))(\alpha'', \beta'', \gamma'').$$

The axis of x' is parallel to the ray

$$\frac{x - \mathrm{A}}{\mathrm{f}} = \frac{y - \mathrm{B}}{\mathrm{g}} = \frac{z - \mathrm{C}}{\mathrm{h}},$$

that is, we have

$$\alpha : \beta : \gamma = \mathrm{f} : \mathrm{g} : \mathrm{h},$$

whence, putting for shortness

$$\Omega = \sqrt{\mathrm{f}^2 + \mathrm{g}^2 + \mathrm{h}^2} \text{ and } \Pi = \sqrt{\mathrm{f}^2 + \mathrm{g}^2},$$

we have

$$\alpha = \frac{f}{\Omega} = \cos N \cos G, \quad \beta = \frac{g}{\Omega} = \cos N \sin G, \quad \gamma = \frac{h}{\Omega} = -\sin N,$$

and thence

$$\tan G = \frac{g}{f}, \quad \sin G = \frac{g}{n}, \quad \cos G = \frac{f}{n}, \quad \cos N = \frac{n}{\Omega},$$

and we thus obtain the values of α', β', γ' ; α'', β'', γ'' in terms of f, g, h and the variable angle H, viz., these are

$$\alpha' = -\frac{g\cos H}{n} + \frac{hf\sin H}{n\Omega}, \qquad \alpha'' = \frac{-g\sin H}{n} - \frac{hf\cos H}{n\Omega},$$

$$\beta' = \frac{f\cos H}{n} + \frac{gh\sin H}{n\Omega}, \qquad \beta'' = \frac{f\sin H}{n} - \frac{hg\cos H}{n\Omega},$$

$$\gamma' = -\frac{n^2\sin H}{n\Omega}, \qquad \gamma'' = \frac{n^2\cos H}{n\Omega},$$

where H is the angular distance of the orbit-pole, along the separator, from the point A. The foregoing values give

$$(a, b, c)(\alpha, \beta, \gamma) = 0,$$

$$(a, b, c)(\alpha', \beta', \gamma') = -\frac{1}{n}\{(ag - bf)\cos H + c\,n\sin H\},$$

$$(f, g, h)(\alpha'', \beta'', \gamma'') = 0,$$

so that the co-ordinates x', y' of the intersection with the ray are given in the form

$$x' : y' : 1 = M : 0 : 0,$$

that is

$$x' = \frac{M}{0} = \infty, \quad y' = \frac{0}{0},$$

but the value of y' is determinate, viz., this is equal to the perpendicular distance of the ray from the point S.

105. In particular when the rays are the special symmetrical system

before considered, then if (a, b, c, f, g, h) refer to the ray 1, we have $f = 0$, $g = 1$, $h = \sqrt{3}$, $\Pi = 1$, $\Omega = 2$, and thence

$$\alpha, \beta, \gamma = 0, \quad \tfrac{1}{2}, \quad \tfrac{1}{2}\sqrt{3},$$

$$\alpha', \beta', \gamma' = -\cos H, \quad \frac{\sqrt{3}}{2}\sin H, \quad -\tfrac{1}{2}\sin H,$$

$$\alpha'', \beta'', \gamma'' = -\sin H, \quad \frac{\sqrt{3}}{2}\cos H, \quad \tfrac{1}{2}\cos H.$$

For the intersection with the ray 1 we have

$$x'_1 = \pm\infty, \; y'_1 = 1,$$

and for the intersections with the other two lines

$$x_2' : y_2' : 1 =$$

$$(3, \sqrt{3}, 2)\left(-\cos H, \frac{\sqrt{3}}{2}\sin H, -\tfrac{1}{2}\sin H\right) = -3\cos H + \tfrac{1}{2}\sin H$$

$$: -(3, \sqrt{3}, 2)\left(0, 1, \frac{\sqrt{3}}{2}\right) \qquad : \frac{3\sqrt{3}}{2}$$

$$: (\sqrt{3}, 1, -2\sqrt{3})\left(-\sin H, -\frac{\sqrt{3}}{2}\cos H, \tfrac{1}{2}\cos H\right) : -\sqrt{3}\sin H - \frac{3\sqrt{3}}{2}\cos H,$$

and

$$x_3' : y_3' : 1 =$$

$$(-3, \sqrt{3}, 2)\left(-\cos H, \frac{\sqrt{3}}{2}\sin H, -\tfrac{1}{2}\sin H\right) = 3\cos H + \tfrac{1}{2}\sin H$$

$$: -(-3, \sqrt{3}, 2)\left(0, \tfrac{1}{2}, \frac{\sqrt{3}}{2}\right) \qquad : \frac{3\sqrt{3}}{2}$$

$$(-\sqrt{3}, 1, -2\sqrt{3})\left(-\sin H, -\frac{\sqrt{3}}{2}\cos H, \tfrac{1}{2}\cos H\right) : \sqrt{3}\sin H - \frac{3\sqrt{3}}{2}\cos H.$$

that is, we have

$$x_2' = \frac{1}{\sqrt{3}}\,\frac{6\cos H - \sin H}{3\cos H + 2\sin H}, \qquad x_3' = -\frac{1}{\sqrt{3}}\,\frac{6\cos H + \sin H}{3\cos H - 2\sin H},$$

$$y_2' = \frac{3}{3\cos H + 2\sin H}, \qquad y_3' = \frac{3}{3\cos H - 2\sin H}.$$

106. Writing herein

$$\cos\omega = \frac{3}{\sqrt{13}}, \quad \sin\omega = \frac{2}{\sqrt{13}}, \quad \tan\omega = \frac{2}{3}, \quad \omega = 33^\circ\,41'$$

the formulæ are readily converted into

$$x_2' = \frac{1}{13\sqrt{3}}\left\{16 - 15\tan(H - \omega)\right\}, \quad x_3' = \frac{1}{13\sqrt{3}}\left\{-16 - 15\tan(H + \omega)\right\}$$

$$y_2' = \frac{3}{\sqrt{13}}\sec(H - \omega), \qquad y_3' = \frac{3}{\sqrt{13}}\sec(H + \omega),$$

where, in regard to this angle ω, it is to be observed that it represents the angular distance from the ecliptic along the separator to a point B, or what is the same thing, the complement of the angular distance on the separator, of the points A and B. We have, in fact, a right-angled spherical triangle Z A B, $\angle Z = 60^\circ$, $\angle A = 90^\circ$, $ZA = 60^\circ$ whence $\sin 60^\circ = \tan AB \cot 60^\circ$, that is, $\tan AB = \sin 60^\circ \tan 60^\circ = \frac{3}{2}$ or $AB = 90^\circ - \omega$.

Hence, $H = \pm 90^\circ$, the orbit-pole is on the ecliptic, $H = \pm(90^\circ - \omega)$, it is at a point B (the intersection of the separator by one of the other two separators), and $H = 0$, it is at the point A on the separator.

The foregoing values of (x_2', y_2') satisfy the equation

$$25y^2 = 39x^2 - 32x\sqrt{3} + 37,$$

and similarly the values of (x_3', y_3') satisfy

$$25y^2 = 39x^2 + 32x\sqrt{3} + 37,$$

which would be useful for the delineation of the planogram.

107. As regards the equation of the orbit we have $x_1' = \pm\infty$, and con-

sequently $x_1' = \pm r_1 = \theta r_1$ if for convenience θ be written to stand for ± 1. The equation of the orbit then is

$$0 = \begin{vmatrix} r, & x', & y', & 1 \\ 1, & \theta, & 0, & 0 \\ r_2(3\cos H + 2\sin H), & \frac{1}{\sqrt{3}}(6\cos H - \sin H), & 3, & 3\cos H + 2\sin H \\ r_3(3\cos H - 2\sin H), & \frac{1}{\sqrt{3}}(-6\cos H - \sin H), & 3, & 3\cos H - 2\sin H \end{vmatrix},$$

that is

$$-(r\theta - x')\,12\sin H =$$

$$y'\left\{\frac{1}{\sqrt{3}}(36\cos^2 H + 4\sin^2 H) - \theta r_2(9\cos^2 H - 4\sin^2 H) + \theta r_3(9\sin^2 H - 4\cos^2 H)\right\}$$

$$-12\sqrt{3}\cos H + 3\theta(3\cos H + 2\sin H)\,r_2 - 3\theta(3\cos H - 2\sin H)\,r_3,$$

where

$$r_2 = \frac{\sqrt{21\cos^2 H - 4\cos H\sin H + \frac{28}{3}\sin^2 H}}{3\cos H + 2\sin H},$$

$$r_3 = \frac{\sqrt{21\cos^2 H + 4\cos H\sin H + \frac{28}{3}\sin^2 H}}{3\cos H - 2\sin H}.$$

Hence, writing $\tan H = \lambda$, and therefore $\sec H = \sqrt{1+\lambda^2}$, which determines the sign of $\sqrt{1+\lambda^2}$, and moreover

$$R_2 = \sqrt{21 - 4\lambda + \frac{28}{3}\lambda^2}, \quad R_3 = \sqrt{21 + 4\lambda + \frac{28}{3}\lambda^2},$$

and thence also

$$(3 + 2\lambda)\,r_2 = R_2, \quad (3 - 2\lambda)\,r_3 = R_3,$$

which last equations, since r_2, r_3 must be positive, determine the signs of the radicals R_2, R_3; the equation of the orbit is

$$r = \theta x' + \frac{y'}{12\lambda\sqrt{1+\lambda^2}}\left\{\frac{\theta}{\sqrt{3}}(36 + 4\lambda^2) - (3 - 2\lambda)R_2 + (3 + 2\lambda)R_3\right\}$$

$$+ \frac{-4\theta\sqrt{3} + R_2 - R_3}{4\lambda},$$

where θ it will be recollected denotes $+1$ or -1 at pleasure.

108. I remark that $\theta = +1$ and $\theta = -1$ may be considered as belonging to positions of the orbit-pole indefinitely near the separator on the opposite sides thereof respectively; the annexed figure represents a portion of the blank sphereogram, and the two sides of the half-separator A′ C′ will be traversed by the orbit-pole, if H extend from 0° to $90° - \omega$ ($= 56° 19'$, value at B′) and thence to 90°, $\theta = +1$ belonging to the side marked + in the figure, and $\theta = -1$ to the opposite side. But the same result may be stated, more conveniently, in reference to the blank sphereogram, as follows:—

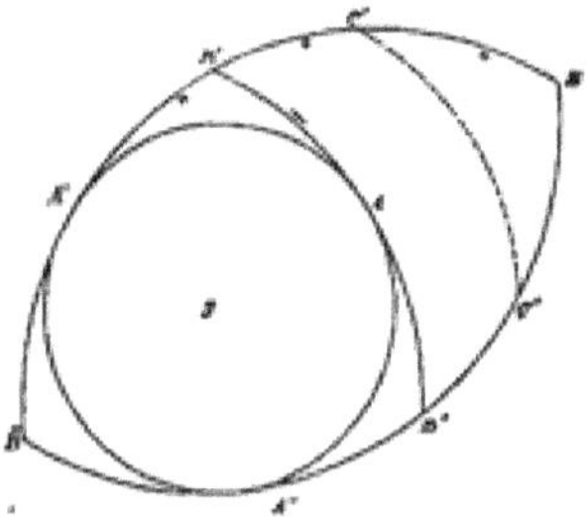

Fig. 10.

H = 0° to H = 56° 19′, $\theta = +1$ belongs to the outside of A B′, viz. to positions within the region of convex orbits,

$\theta = -1$, to inside of A B′,

H = 56° 19′ to H = 90°, $\theta = +1$ belongs to inside of B′ C′.

H = 90° to H = 123° 41′, $\theta = +1$ belongs to inside of C′ B,

the last-mentioned values being identical with those for H = 90° to H = 56° 19′, $\theta = -1$: viz. the formula for $H = 90° + K$, $\theta = +1$ is equivalent to that for $H = 90° - K$, $\theta = -1$.

109. I consider some particular cases.

Orbit-pole at A: here H = 0 and therefore $\lambda = 0$, $R_1 = R_3 = \sqrt{21}$; the orbit is $r = \theta x' + \frac{\theta\sqrt{3}}{\lambda}(y' - 1)$, viz. it is the right line $y' - 1 = 0$.

Orbit-pole in the neighbourhood of B. Suppose first $H = 90° - \omega - \epsilon$, $\lambda = \cot\omega - \epsilon \operatorname{cosec}^2\omega = \frac{3}{2} - \frac{13}{4}\epsilon$, $3 - 2\lambda_1 = \frac{13}{2}\epsilon$, is positive, and therefore R_2 is positive, and we have $R_2 = 6$, $R_3 = 4\sqrt{3}$; whence the equation is

$$r = \theta x' + \frac{y'}{9\sqrt{13}}(15\theta + 24) + 1 - \frac{2}{\sqrt{3}}(\theta + 1),$$

viz. $\theta = -1$, this is

$$r = +x' + \sqrt{\frac{1}{13}}\,y' + 1,$$

and $\theta = +1$, it is

$$r = x' + \sqrt{\frac{13}{3}}\,y' + 1 - \frac{4}{\sqrt{3}},$$

and so secondly, if $H = 90^\circ - \omega + \epsilon$, $\lambda = \frac{3}{2} + \frac{13}{4}\epsilon$, $3 - 2\lambda$, $= -\frac{13}{2}\epsilon$, is negative, or R_y is also negative, viz. $R_x = 6$, $R_y = -4\sqrt{3}$, and the equation is

$$r = \theta x' - \frac{y'}{9\sqrt{3}}(15\,\theta - 24) + 1 - \frac{2}{\sqrt{3}}(\theta - 1),$$

viz. $\theta = +1$, this is

$$r = x' - \sqrt{\frac{1}{13}}\,y' + 1,$$

and $\theta = -1$, it is

$$r = -x' - \sqrt{\frac{13}{3}}\,y' + 1 + \frac{4}{\sqrt{3}}.$$

At the point B there are thus four orbits: viz. $H = -90^\circ - \omega - \epsilon$, $\theta = +1$, and $H = 90^\circ - \omega + \epsilon$, $\theta = -1$, these are orbits wherein the excentricity is $= \sqrt{\frac{16}{3}}$, $= 2·309$, agreeing with that found for the point B in planogram No. 1, or say for an orbit pole near B in the direction of the meridian; whereas for $H = 90^\circ - \omega - \epsilon$, $\theta = -1$ and $H = 90^\circ - \omega + \epsilon$, $\theta = +1$ the excentricity is $\sqrt{\frac{16}{13}} = 1·101$.

Suppose again that the orbit-pole is on the ecliptic, or say $H = 90^\circ - \epsilon$, $\lambda = +\infty$, $R_x = 2\sqrt{\frac{7}{3}}\,\lambda$, $R_y = -2\sqrt{\frac{7}{3}}\,\lambda$, and $\sqrt{1 + \lambda^2} = \lambda$, and the equation is

$$r = \theta\left(x' + \frac{y'}{3\sqrt{3}}\right) + \sqrt{\frac{7}{3}},$$

and similarly for $H = 90^\circ + \epsilon$, $\lambda = -\infty$, $R_x = 2\sqrt{\frac{7}{3}}\,\lambda$, $R_y = -2\sqrt{\frac{7}{3}}\,\lambda$, $\sqrt{1 + \lambda^2} = \lambda$, and the equation still is

$$r = \theta\left(x' + \frac{y'}{3\sqrt{3}}\right) + \sqrt{\frac{7}{3}},$$

viz. θ retaining the same sign, there is no discontinuity in the passage through 90°.

The excentricity, whether $\theta = +1$ or $= -1$, is $\sqrt{\frac{28}{27}}$, $= 1{\cdot}018$, agreeing with Planogram No. 2.

110. For the more complete discussion of the excentricity, we have

$$e^2 = 1 + \frac{1}{144\lambda^4(1+\lambda^2)}\left(\frac{4\theta}{\sqrt{3}}(9+\lambda^2) - (3-2\lambda)R_4 + (3+2\lambda)R_1\right)^2.$$

The excentricity cannot be less than 1, which is evidently right, for the point 3 being at infinity, the orbit cannot be an eclipse. We may have $e = 1$ (or the orbit a parabola), viz. this will be the case if

$$\frac{4\theta}{\sqrt{3}}(9+\lambda^2) - (3-2\lambda)R_4 + (3+2\lambda)R_1 = 0.$$

Proceeding to rationalize this equation, we have first

$$(3-2\lambda)^2 R_4^2 + (3+2\lambda)^2 R_1^2 - \frac{16}{3}(9+\lambda^2)^2 = 2(9-4\lambda^2)R_4R_1,$$

viz. substituting for R_4, R_1 their values $\sqrt{21 + 4\lambda + \frac{28}{3}\lambda^2}$ and $\sqrt{21 + 4\lambda + \frac{28}{3}\lambda^2}$, this is found to be

$$2(9-4\lambda^2)\sqrt{\left(21+\frac{28}{3}\lambda^2\right)^2 - 16\lambda^2} = -54 + 336\lambda^2 + \frac{208}{3}\lambda^4;$$

or, what is the same thing,

$$(9-4\lambda^2)\sqrt{3969 + 3384\lambda^2 + 784\lambda^4} = -81 + 504\lambda^2 + 104\lambda^4,$$

whence, squaring and reducing, we have

$$432(4\lambda^8 - 248\lambda^6 - 819\lambda^4 + 162\lambda^2 + 729) = 0;$$

or, what is the same thing,

$$432(\lambda^2+1)(4\lambda^6 - 252\lambda^4 - 567\lambda^2 + 729) = 0,$$

or, finally, the condition for a parabola is

$$4\lambda^6 - 252\lambda^4 - 567\lambda^2 + 729 = 0.$$

111. I stop to remark that this equation may be obtained differently, as follows. Since the point 1 is at infinity on the axis of x, this line will be the axis of the parabola; or the equation of the parabola will be

$$-y^2 + 4ax + 4a^2 = 0,$$

and we have therefore

$$-y_4^2 + 4ax_4 + 4a^2 = 0,$$
$$-y_3^2 + 4ax_3 + 4a^2 = 0,$$

that is

$$1 : 4a : 4a^2 = x_4 - x_3 : y_4^2 - y_3^2 : -y_4^2 x_3 + y_3^2 x_4,$$

and therefore

$$(y_4^2 - y_3^2)^2 = -4(x_4 - x_3)(y_4^2 x_3 - y_3^2 x_4),$$

as the condition for a parabola.

But the values of x_4, y_4; x_3, y_3, *ante* No. 104, introducing λ in the place of II, are

$$x_4 = \frac{1}{\sqrt{3}}\,\frac{6-\lambda}{3+2\lambda}, \quad x_3 = \frac{1}{\sqrt{3}}\,\frac{6+\lambda}{3-2\lambda},$$

$$y_4 = \frac{3\sqrt{1+\lambda^2}}{3+2\lambda}, \quad y_3 = \frac{3\sqrt{1+\lambda^2}}{3-2\lambda},$$

and thence

$$x_4 - x_3 = \frac{4}{\sqrt{3}}\,\frac{9+\lambda^2}{9-4\lambda^2},$$

$$y^2_4 - y^2_3 = -\frac{216\lambda(1+\lambda^2)}{(9-4\lambda^2)^2},$$

$$y^2_4 x_3 - y^2_3 x_4 = -\frac{36}{\sqrt{3}}\,\frac{(1+\lambda^2)(9-\lambda^2)}{(9-4\lambda^2)^2},$$

and substituting these values and omitting a factor $\frac{1+\lambda^2}{(9-4\lambda^2)^3}$, the result is

$$\frac{243\lambda^2(1+\lambda^2)}{9-4\lambda^2} = (9+\lambda^2)(9-\lambda^2).$$

viz. this is

$$(4\lambda^2 - 9)(\lambda^4 - 81) - 243\lambda^2(\lambda^2 + 1) = 0,$$

that is

$$4\lambda^6 - 252\lambda^4 - 567\lambda^2 + 729 = 0,$$

as before.

112. The equation considered as a cubic equation in λ^2 has its three roots real, but only two of them are positive; viz. there is a root not very different from 1, and which is easily approximated to by writing $\lambda^2 = 1 - x$, this gives

$$4x^3 + 240x^2 - 1068x + 86 = 0,$$

or nearly $x = \frac{86}{1068} = \cdot 08$; a second approximation gives $x = \cdot 0802$; or we have $\lambda^2 = \cdot 9198$, $\lambda = \cdot 9592$, whence $H = 43^\circ\, 49'$. Substituting in the equation

$$\frac{4\theta}{\sqrt{3}}(9 + 4\lambda^2) - (3 - 2\lambda)R_2 + (3 + 2\lambda)R_1 = 0,$$

this will be satisfied by $\theta = -1$, viz. the parabola belongs (as it obviously should do) to a point of A B′ within the triangle B B′ B″.

To obtain the other positive root we may write the equation in the form

$$\lambda^2 = 63 + \frac{141{\cdot}75}{\lambda^2} - \frac{182{\cdot}25}{\lambda^4},$$

the approximate value $\lambda^2 = 63$, gives more nearly $\lambda^2 = 65$ and then

$$\lambda^2 = 63 + \frac{141{\cdot}75}{65} - \frac{128{\cdot}24}{4225}, = 65{\cdot}177,$$

whence $\lambda = 8{\cdot}073$ or $H = 82^\circ\, 56'$. Substituting in the equation

$$\frac{4\theta}{\sqrt{3}}(9 + 4\lambda^2) - (3 - 2\lambda)R_2 + (3 + 2\lambda)R_1 = 0,$$

we have $\theta = +1$, viz. this parabola belongs to a point of B′ C′ within the triangle B B′ B″.

The two values of e for $\theta = +1$ and $\theta = -1$, are each infinite for $\lambda = 0$, and they become equal for $\lambda = \infty$ (viz. when the orbit-pole is on the ecliptic), but not in any other case; in fact they can only do so for $9 + \lambda^2 = 0$, or else for $(3 - 2\lambda) R_2 = (3 + 2\lambda) R_1$, that is, $\lambda (288 + 128 \lambda^2) = 0$, viz., $\lambda (9 + 4\lambda^2) = 0$.

113. In further explanation I give a diagram of the excentricity.

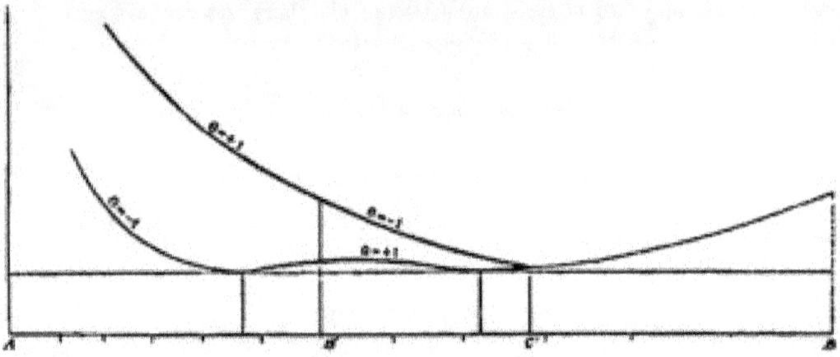

Fig. 11.

The base A B′ C′ B is here the broken line A B′ C′ B′ of figure 10: the ordinates along the base A C′ (= 90°) of the two continuous curves exhibit the values of e, as given by $\theta = +1$ and $\theta = -1$ respectively; the dotted curve on the base C′ B (= C′ B′) is merely the upper curve on the base C′ B′ transferred to the base C′ B′; and the curve composed of the lower curve on the base A C′ and of the dotted curve gives by its ordinates the value of the excentricity as the orbit-pole moves along A B′ B within the triangle B′ B B″: the upper curve on the base A B′ gives by its ordinates the value of the excentricity as the orbit-pole moves along A B′ on the other side thereof, that is, within the convex region.

The base of the diagram is graduated not for the value of H, but for that of the angular distance (or distance in longitude) of the orbit-pole from the point A (or A′); viz. this is the angle opposite H in a right-angled spherical triangle, the sides and hypothenuse of which are 60°, H, c; writing β for the angle in question we have

$$\cos \epsilon = \frac{1}{2} \cos H, \quad \tan \beta = \frac{2}{\sqrt{3}} \tan H \cdot \left(= \frac{2\lambda}{\sqrt{3}}\right),$$

and any position of the orbit-pole on the separator may be conveniently laid down by means of this angle β. The values of β corresponding to the before-mentioned values $\lambda = \cdot 9592$ and $\lambda = 8 \cdot 073$ are $\beta = 47^\circ\ 54'$ and $\beta = 83^\circ\ 53'$ respectively.

Art. Nos. 114 and 115. The Spherogram and Isoparametric Lines—General Considerations.

114. We first construct a blank spherogram, as already explained (and see also Plates IV. and V.), viz., we draw on the stereographic projection a hemisphere—say the northern hemisphere: the meridians being radii and the parallels of colatitude circles with the pole as centre; the parallel of 60° is the regulator circle, and the separators are great circles touching this at the points A, A, A, in longitudes 30°, 150°, 270° respectively; the separators intersect in the points B, B, B, in the northern hemisphere, and they are produced to meet again in the points B, B, B, of the southern hemisphere; but instead of taking the whole northern hemisphere, we omit portions thereof, and take in the opposite portions of the southern hemisphere; the spherogram being thus bounded by portions of the separator circles, and consisting of the inner spherical triangle B, B, B, and three surrounding triangles B, B, B. The inner triangle contains the regulator-circle, touching its sides at the points A, A, A respectively, and dividing it into an inner circular region and three surrounding regions A, B, A; these last are the *loci in quibus* of the orbit-poles which correspond to convex orbits; and to mark them off from the other regions, it is proper to shade them in the spherogram. Excluding them from consideration, we have the inner circular region and the outer triangular regions separated off from each other by the shaded regions, except at the points A, where these are thinned away to nothing. The points A are positions of the orbit-pole for which the orbit is indeterminate; and consequently any parameter belonging to the orbit is also indeterminate. Hence the isoparametric line for any given value of the parameter will always pass through the points A; that is, all the isoparametric lines will pass through these points, which are thus points of connexion between

the inner circular region and the three outer regions, but it must be recollected that for certain given values of the parameter, the points A may be isolated points on the isoparametric line.

115. It is sometimes necessary (more particularly as regards the Time-spherogram and isochronic lines) to distinguish from each other the several points A and B; and for this purpose I consider the several points, as situated in the spherogram, to be accented in the following manner:—

$$\begin{array}{ccccc} B^{iv} & & B' & & B \\ & A' & & A & \\ & B''' & A'' & B'' & \\ & & & & \\ & & B^{v} & & \end{array}$$

so that the inner triangle is $B'\,B''\,B'''$ and the outer triangles are $B\,B'\,B''$, $B'\,B^{iv}\,B'''$ and $B'''\,B^{v}\,B''$ respectively; this distinction has been already partially made in Fig. 10.

Articles Nos. 116 to 122. The e-spherogram and Iseccentric Lines. See Plate IV.

116. Constructing a blank spherogram as above, we may from the tables for planograms Nos. 1 and 2 lay down numerically the values of the excentricity at the several points of each meridian for the longitudes 0°, 30°, . . 330°, viz.

LONGITUDES 0°, 60°, 120°, 180°, 240°, 300°.

Planogram No. 2 shows that e increases from 0 at the centre to ∞ at 60°, then, 60° to 63° 26′ (shaded region), it diminishes from ∞ to 4·912; on passing 63° 26′ it changes abruptly to 1·853; thence diminishes to a minimum = ·628 at 59°, and again increases to 1·018 at 90°.

LONGITUDES 10°, 210°, 330°.

Planogram No. 1, part 1, shows that e increases from 0 at the centre to ∞ at 60°, then, 60° to 73° 54′ (shaded region), it diminishes from ∞ to 2·309, this last value being at a point $B_{,}$ the termination of the spherogram.

LONGITUDES 30°, 150°, 210°. Planogram No. 1, part 2, and for values over 90°, part 1, shows that e increases from 0 at the centre to ·264 at 60° (point A), ·869 at 90°, and 2·309 at 102° 6′, point B.

It will be recollected that, although e has the same value, 2·309 at the two opposite points B, yet there is an abrupt change of orbit, indicated by the change of sign of A (= ± e).

117. Planogram No. 3 shows the directions at the points A of the several isocentric lines. Planogram No. 4, if the calculations were completed, would give the value of the excentricity at the several points of the ecliptic, but besides the already-mentioned values 1·018 at 0°, 60°, &c., and ·868 at 30°, 90°, &c., the only value calculated is ·906 at 15°, 45°, &c. It thus appears that the excentricity = 1·018 for longitude 0° diminishes through ·906 at 15° to ·868 at 30°, and then again increases through ·906 at 45° to 1·018 at 60°, and so on through successive intervals of 60°.

118. Planogram No. 5, if the calculations were completed, would give the value of e for the arc AB within the shaded region (but no values have been found except those given by Planograms 1 and 2, viz. $e = \infty$ at A, = 4·912 at longitude 30° from A, and = 2·309 at B); and it would also give the value of e for the whole bounding arc ABB within the exterior triangular region. We have $e = \infty$ at A, = 1·853 at longitude 30° from A, = 1 at distance H = 43° 49′ from A, = 1·101 at B, and then proceeding along the arc BB, = 1 at distance H = 82° 56′ from A, = 1·018 on the ecliptic, and, finally, = 2·309 at B. The two values $e = 1$ are very important, as will presently appear, with regard to the parabolic curve.

119. It is now easy to trace the form of the isocentric lines.

$e = 0$, the curve is a point at the centre, and for any value less than ·264 it is a trigonoid form surrounding the centre, the maxima radii being directed towards the points A. The points A belong as isolated points to all these curves.

$e = ·264$, the curve is tricuspidal, having a cusp at each of the points A. The numerical values seem to show a singularly blunt form of cusp (the points A are, in fact, not ordinary cusps, but singular points of a higher order); but the data do not enable me to draw with certainty the precise forms of the arcs between the three cusps: the wavy form was drawn purposely, but there is no sufficient evidence for its correctness.

120. It is convenient to pass at once to the case $e = 1$, or say the parabolic curve, locus of the orbit-pole when the orbit is a parabola. This is a three-looped curve cutting itself (having a node) at each of the points A; and it appears from planogram No. 5 that each loop touches at four points (two points, $H = 43° 49'$, and two points, $H = 82° 56'$), the sides of the bounding triangle B B B. The loop thus divides the triangle B B B into six regions, viz. one within the loop, two subjacent, two lateral, and one superjacent.

For any value between $e = \cdot 264$ and $e = 1$, the curve is a three-looped curve intersecting itself at the points A, and such that the loops lie wholly within those of the parabolic curve, and the remaining portions between the parabolic and cuspidal curves.

121. For any value of $e > 1$, we must imagine a three-looped curve intersecting itself at the points A, the loops respectively containing those of the parabolic curve, and the remaining portions within the regulator-circle lying between the regulator-circle and the parabolic curve; and we must then obliterate so much of each loop as lies in the shaded regions, or outside the spherogram; viz. instead of a continuous loop there will be thus a broken loop with detached portions thereof in the subjacent regions, the lateral regions, and the superjacent region respectively. More precisely this is the form for any value of e from $e = 1$ to $e = 1\cdot 101$, but for this last value the unobliterated portion for each lateral region evanesces; for any value of e between $e = 1\cdot 101$ and $e = 2\cdot 309$, the unobliterated portions lie wholly within the subjacent regions and the superjacent region; for $e = 2\cdot 309$ the portion within the superjacent region evanesces; and for any greater value of e the unobliterated portion lies wholly within the subjacent regions, the loop being thus a mere fragment.

122. The isoeccentric curves within the shaded regions form a distinct system: such curves belong to the values $e = 2\cdot 309$ to $e = \infty$, and any one of them is a fragment of a three-looped curve intersecting itself at the points A, obtained by obliterating so much of the complete curve as lies outside the shaded regions. But it is perhaps better to disregard these curves altogether, thus in effect excluding the shaded regions from the spherogram.

Article Nos. 123 to 143. The Time-spherogram and Isochronic Lines. See Plate V.

123. We construct a blank spherogram, and lay down upon it the parabolic curve; we may then lay down (as will be explained) the numerical values, say of the times T_{12}, but in order to gain some idea of the form of the T_{12}-lines I will first consider the question in a more general manner.

124. When the orbit is a line, parabola, or hyperbola, we may distinguish it by the letters L, P, H accordingly; and by the numbers 1, 2, 3, written in the proper order, show the arrangement of the three points on the orbit; observe that if 1 be the middle point on the orbit, we may write indifferently 213 or 312, and so in other cases, the fixation of the middle number is alone material. When the orbit is a line the distances of the points are always finite; and if the orbit be, for example, L 123, then T_{12} and T_{23} are each = 0, but T_{31} is non-existent. For the parabola and hyperbola the distances are in general finite; but it is necessary to distinguish for the parabola, *e.g.* the case P$\dot{1}$23 where an extreme point, and for the hyperbola, *e.g.* the cases H $\dot{1}$23 and H $\dot{1}2\dot{3}$ where one or each of the extreme points is, at infinity. We have in these cases respectively

P 123,	T_{12} finite,	T_{23} finite,	$T_{31} = \infty$
P $\dot{1}$23.	$T_{12} = \infty$,	T_{23} finite,	$T_{31} = \infty$

and it may be added, as regards P $\dot{1}$23, that, by a continuous change of the parabolic orbit the point 1 may change over to infinity on the other half-branch of the parabola, or the arrangement become P 23$\dot{1}$. And, moreover

H 123,	T_{12} finite,	T_{23} finite,	T_{31} non-existent.
H $\dot{1}$23,	$T_{12} = \infty$,	T_{23} finite,	T_{31} non-existent.
H $\dot{1}2\dot{3}$,	$T_{12} = \infty$,	$T_{23} = \infty$,	T_{31} non-existent.

Thus the proper symbol L 123, P $\dot{1}$23, &c. as the case may be, will always at once indicate as to each of the times T_{12}, T_{23}, T_{31}, whether this is = 0, finite, infinite, or non-existent.

125. We may without difficulty attach to the several portions of the regulator, the separators and the parabolic curve, to each portion its proper symbol L, P, H and 123, $\dot{1}$23, &c. as the case may be.

First, as to the regulator, it is obvious that this is separated by the points A into the three portions L 213, L 321, L 132, respectively. And inside the regulator, adjacent to these, we have portions of the parabolic curve P 213, P 321, P 132, respectively.

Again, for one of the separators, say $B^{iv}\,B'\,A\,B'''\,B^{v}$ (see here and in all that follows the notation-diagram, No. 115); since the point 2 is here at infinity this must be at every portion thereof either H $13\dot{2}$ or else H $3\dot{1}2$. The point B^{iv} is H $\dot{1}32$ and the point B^{v} is H $3\dot{1}2$; consequently, as the orbit-pole passes along the separator from B^{iv} to B^{v}, the symbol is at first H $13\dot{2}$ and at last H $3\dot{1}2$; the transition takes place at the point of contact of the parabolic curve which is indifferently P $13\dot{2}$ or P $\dot{2}13$. (In further explanation of the transition, consider the orbit-pole as passing from B^{iv} to B^{v}, not on the separator, but indefinitely near it; it can only do so by twice crossing the parabolic curve near the point of contact; the orbit is first H $13\dot{2}$, or say H 132, then P 132, then an ellipse, which when the orbit-pole again arrives at the parabolic curve changes into P 312; and it finally becomes H 312 or H $3\dot{1}2$.)

126. Again, since, on the two separators through B^{iv}, in the portions adjacent to B^{iv}, the symbols are H $13\dot{2}$ and H $\dot{1}32$, it is clear that in the adjacent portion of the parabolic curve (terminated each way by a point of contact with these separators respectively) the symbol must be P 132; at the point of contact with the first-mentioned separator $B^{iv}\,B'\,A\,B'''\,B^{v}$, this becomes P $13\dot{2}$, = P $\dot{2}13$; and beyond the point of contact it becomes P 213, continuing so until it arrives at the next point of contact with the separator $B'\,A'\,B''$: there is always in the symbol for the parabolic curve this change of form as we pass through a point of contact with a separator; and there is the same change, when *travelling along the loop* (that is without going inside the regulator) we pass through a point A. The foregoing considerations fully explain how the proper symbol is to be attached to each portion of the regulator, the separators, and the parabolic curve: to avoid confusion, I have abstained from attaching them in the Plate.

127. Imagining the symbols attached as above, it at once appears that, for the two portions $A'\,A$ and $A\,A''$ of the regulator curve, we have $T_{,,} = 0$; while, for the arc $A''\,A'$ of the parabolic curve we have $T_{,,} = \infty$. Moreover, $T_{,,}$ can only be infinite on one of the separators through B''' and on the parabolic curve; and the symbols show that the curve $T_{,,}$ is made

up, in a peculiar discontinuous manner, of portions of these two separators and of the parabolic curve, as shown by the strongly marked line of the figure; we have thus the boundary of certain *lightly* shaded regions within which (as well as within the shaded regions) T_{12} is non-existent; excluding these, the remaining regions (instead of a trilateral symmetry) have a symmetry about the axis B B‴; there are still four regions which may be distinguished as the inner region, the axial outer region, and the lateral outer regions; or, more shortly, as the inner, axial, and lateral regions.

128. The times T_{12}, T_{23}, T_{31} are calculated, Planogram 1, part 1, for the meridian long. 90°, and ditto part 2 for the meridian long. 270°; and in Planogram 2 for the meridian long. 180°. As regards these last values, it is easy to see that, in order to pass to the meridian long. 0°, the numbers 2, 3 must be interchanged; that is, long. 0°, the T_{12}, T_{23}, T_{31} are respectively equal to the values, long. 180°, T_{13}, T_{32}, T_{21}. Moreover, the numbers 1, 2, 3 may be changed into 2, 3, 1, or into 3, 1, 2, provided the longitude is increased by 120° and 240° in the two cases respectively; that is,

$$\begin{aligned} T_{12}, \text{ long. } \omega &= T_{21}, \text{ long. } \omega \\ &= T_{31}, \text{ long. } (\omega + 120°) \\ &= T_{23}, \text{ long. } (\omega + 240°) \end{aligned}$$

129. By means of the foregoing two relations, T_{12} for the several longitudes 0°, 30°, 60°... 330°, is given as equal to the T_{12}, T_{23}, or T_{31}, for long. 90°, 270°, or 180°, that is, to the T_{12}, T_{23}, or T_{31}, of Planogram No. 1, part 1 or 2, or of Planogram No. 2. For example, T_{12} long. 240° = T_{31} long. 0° = T_{21} long. 180°, that is, it is equal to the T_{21} of Planogram No. 2. We thus find

Long.	T_{12} is =		
0°	T_{13} of Plan. No. 2		
30°		T_{23} of Plan. No. 1, pt. 2	
60°	T_{13} ,,		
90°			T_{12} of Plan. No. 1, pt. 1
120°	T_{23} ,,		
150°		T_{31} ,,	
180°	T_{12} ,,		
210°			T_{23} ,,
240°	T_{21} ,,		
270°		T_{12} ,,	
300°	T_{23} ,,		
330°			T_{31} ,,

and observing that for Planogram No. 1, part 1 or 2, we have $T_{14} = T_{11}$, it hence appears as above, that the meridian 30°–210° is an axis of symmetry of the spherogram. In what precedes it has been assumed that the colatitudes only extend from 0° to 90°, but in the spherogram they extend for the meridians 30°, 150°, 270°, to the colatitude 106° 6′, the values for the colatitudes above 90° are those for the omitted portions 90° to 73° 54′ of the opposite meridian.

N.B. A meridian extends from the pole *in one direction only*, unless the contrary is expressed or implied, as in speaking of a meridian 0°–180°.

130. I attend, in the first instance, to the axis of symmetry or meridian 30°–210°. Proceeding along the meridian long. 30° or towards the point A, the value of T_{11} decreases from 1 at the centre to a minimum = ·950 at colatitude 11° (call this the point X), and it then increases to 1·983 at A, and thence to 58·62 at 90° and ∞ at the parabolic boundary of the axial region. In the opposite direction it increases from 1 at the centre to ∞ at the parabolic boundary of the inner region. The minimum value ·950 on the axis of symmetry indicates a node on the isochronic curve; that is, the point X is a node on the isochronic T_{11} = ·950. This will consist of two branches, proceeding from A′, A″, respectively, cutting the axis and each other at X, then again cutting at A, and thence passing on into the axial region, and respectively terminating on the separator boundary B′ A B″ thereof.

131. This curve, which I call the nodal isochronic, divides the inner region into a loop, anti-loop, and two side regions. On each of the meridians 0°, 60°, the value of T_{11} diminishes from 1 at the centre to a minimum which is less, and then increases to a maximum which is greater, than ·950; the value then diminishes to 0 on the regulator: on emergence of the meridian from the shaded into the axial region, the value is = ·909, and it thence increases to ∞ at the parabolic boundary of the axial region; these data further determine the form of the nodal isochronic, viz., each of the two half meridians cuts the loop twice, and again cuts the curve in the axial region.

The nodal isochronic, at each of the points A′, A″, continues its course into the lateral region, returning to the same point A or A′, so as to form in each of the lateral regions a loop. Considering the loop as formed of two branches, each proceeding from A′ or A″, the one which is the continuation of the course within the inner region I call the lower branch; the other, the upper branch; and I say that the upper branch *touches* the separator at A′ or A″. The two branches and the entire loop lie on the left-hand side (or side away from A) of the meridians through A′ or A″. As to the contact of the upper branch of this and other isochronics at A′ or A″ with the separator, see *post* No. 142.

132. It is convenient at this point to consider the form of the isochronic curves within the axial region. The parabolic boundary thereof is an isochronic $T_{11} = \infty$, and it thence appears that for any large value of T_{11} the isochronic curve (portion of the curve) is a curve not meeting the parabolic boundary, and terminated each way in the separator boundary B′ A B″. As the value of T_{11} diminishes, the curve (which is of course always symmetrical in regard to the axis) bends inwards towards the point A and for $T_{11} = 1{\cdot}983$ (value on the axis at A) the curve acquires a cusp at A. I call this the cuspidal isochronic; I remark that it intersects in the axial region each of the meridians 0° and 60°.

As T_{11} further diminishes to any value between 1·983 and ·950, the curve, commencing in the separator boundary, passes through A into the inner region, and, forming a loop within the loop of the nodal isochronic, emerges through A into the axial region, terminating again in the separator boundary.

133. On the meridians 90°, 330°, through the points B′, B″, respectively, the value of T_{11} diminishes from 1 at the centre to 0 at the regulator, where these meridians are considered as terminating.

On the meridians 120°, 300° (meridian at right angles to the axis of symmetry), the value of T_{11} diminishes from 1 at the centre to a minimum less than ·878, and then increasing to a maximum of over ·895 diminishes to 0 at the regulator. On emergence of the meridian from the shaded and half-shaded region on the parabolic boundary of the lateral region the value is $= \infty$, and it thence diminishes to 1·148 on the separator boundary B″ B′ or B′ B″.

On the meridians 150°, 270°, which pass through A′, A″, respectively, the value of T_{11} increases from 1 at the centre to 1·377 at the regulator, and thence through 2·255 at 90° to ∞ at B″ or B′.

And finally, on the meridians 180°, 240°, the value of T_{11} increases from 1 at the centre to ∞ at the parabolic inner boundary, and then on emergence from the half-shaded and shaded region at the separator boundary $B'' A'$ or $B''' A''$, the value is $= \infty$, and it thence diminishes to a minimum under 6·343, and again increases to ∞ at the separator boundary $B''' B^{iv}$ or $B''' B^{v}$.

134. By what precedes, it appears that on the separator boundary $B^{iv} B'$ or $B^{v} B''$ of either of the lateral regions, the values of T_{13} is each extremity $= \infty$, and at an intermediate point $= 1·148$; there is consequently a minimum value less than 1·148, and therefore two points at each of which the value is $= 1·983$.

Now resuming the consideration of the cuspidal isochronic ($T_{13} = 1·983$) as regards the remaining portions thereof, viz., those in the lateral and inner regions; and considering first the lateral region $B''' B^{iv} B'$, there will be from each of the points just referred to on the boundary $B^{iv} B'$ a branch, one (which I call the lower branch) from the point nearer B', passes, on the right-hand side of the meridian through A', to A'; the other (which I call the upper branch) proceeding from the point nearer B^{iv}, cuts the same meridian, and then on the left-hand side thereof arrives at A', touching there the separator: at A'' in the other lateral region there are in like manner an upper and a lower branch (situate symmetrically, in regard to the axis, with the upper and lower branches at A'); and continuous with the two lower branches there is a branch from A' to A'', through the anti-loop of the inner region.

135. Imagine the given value of T_{13} as continuously increasing from the value ·950, which belongs to the nodal isochronic; and attend in the first instance to the form within the lateral regions. There will be a loop of continually increasing magnitude (viz., the loop for a larger value of T_{13} will always wholly include that for a smaller value); each loop formed by an upper branch, which at A' touches the separator, and a lower branch the direction of which from A' is variable. So long as T_{13} is less than 1·377 (value at A' along the meridian) the lower branch, and consequently the whole loop, will lie on the left hand of the meridian; but when T_{13} is $= 1·377$, the lower branch touches the meridian, and for any greater value of T_{13} lies on the right of the meridian; and in either of the last-mentioned cases the loop is cut by the meridian, and thus lies partly on the left, and partly on the right thereof.

136. Now by what precedes there is on the separator boundary $B' B^{iv}$ of

the lateral region a point where T_{11} has a minimum value less than 1·148, and consequently, for any given value, say for a value between this minimum and 1·377, there are on B′ B″ two points where T_{11} has the given value. These points cannot lie on the loop of the curve belonging to the given value (for this loop is wholly on the left hand of the meridian); hence the complete curve for the given value of T_{11} will include (within the lateral region) besides the loop, a branch uniting the two points in question; say a link branch.

137. It follows that there is between T_{11} = 1·377 and 1·983, a value (to fix the ideas, say = 1·82?, it being understood that I do not attempt to determine this value) for which the loop and link branch will unite themselves together, the point of junction becoming as usual a node; viz., there will be a curve T_{11} = 1·82? having in the two lateral regions respectively the nodes Y, Y′; or say the curve has in each lateral region a self-intersecting loop. For any greater value of T_{11} (as for example the value 1·983 belong to the cuspidal curve) there are two branches inclosing the self-intersecting loop; for a less value, as has been seen, instead of the self-intersecting loop, there is a loop and link branch; at least this is the case until for the minimum value < 1·148 of T_{11} on the separator boundary B″ B′ the link branch disappears. For smaller values down to T_{11} = ·950, which belongs to the nodal isochronic, there is no link branch, but only the loop; and as T_{11} diminishes below this value, there is still a continually diminishing loop, lying wholly on the left hand of the meridian, and with its upper branch always touching the separator; and ultimately for T_{11} = 0 the loop vanishes.

138. We have attended wholly to the lateral regions; but the consideration of the axial and inner regions is very easy: for any value between the values 1·983 and ·950, there are in the axial region (between the nodal and cuspidal curves) two branches each proceeding from the separator to A, where they unite, and, crossing each other, pass into the inner region, forming a loop within the loop of the nodal isochronic; and, moreover, there is in the inner region a branch, the continuation of the lower branches of the lateral loops, uniting the points A′, A″, and lying between the nodal and cuspidal isochronic. And for T_{11} less than ·950 there are in the axial region between the nodal curve and the separator, two branches, each proceeding from the separator to A, where, crossing

each other, they enter the inner region passing outside the nodal curve (or in the side regions of the inner region) to the points A′, A″, where they respectively join on to the lower branches of the lateral loops. Ultimately, for $T_{11} = 0$, the curve coincides with the finite portions A A′, A A″ of the regulator circle.

139. We have finally to consider the case T_{11} greater than 1·983: there is in the axial region a branch lying outside the cuspidal curve, and extending from separator to separator; in each lateral region two branches (lying outside those of the cuspidal curve) each proceeding from A′ (or A″) to the separator boundary B′ B″ or B″ B′, an upper branch touching the separator at A′ or A″, and a lower branch; and in the inner region, a branch (continuation of the lower branches) lying between the cuspidal curve and the parabolic boundary of the anti-loop, and uniting the points A′, A″. In the ultimate case $T_{11} = \infty$, the curve coincides with the before-mentioned discontinuous curve composed of portions of the parabolic curve and of two separators.

140. To obtain a comprehensive statement of the foregoing results, we may (as in the case of the isoeccentric lines) imagine the curves completed and rendered continuous by the insertion of portions lying outside the spherogram, or within the half-shaded and shaded regions; which inserted portions are to be ultimately obliterated. The upper and under branches terminating in the separator boundary of a lateral region are thus completed into a loop; the link branch into a closed curve or oval; the vanishing of the link branch happens when the oval, on the point of passing outside the separator boundary of the lateral region, just touches this boundary; as T_{11} diminishes to the value for which this happens, and continues still further to diminish, I think it may be assumed that there is some value (to fix the ideas, say $T_{11} = 1·10$?, but I do not attempt to determine it) for which the oval becomes a conjugate point, viz., for this value $T_{11} = 1·10$? the curve will have two conjugate points (nodes) Z′, Z″, outside the two lateral regions respectively.

141. We may now state the forms of the curve. The points A, A′, A″, are always nodes, viz., A′, A″, nodes with real branches, but A is either a conjugate point, a cusp, or a node with real branches.

$T_{11} > 1·983$: two looped curve, containing within it A as a conjugate point:

as T_{11} diminishes, the curve bends inwards towards A, and

$T_{11} = 1·983$: cuspidal isochronic ; A, a cusp.

T < 1·983, the curve cuts itself at A, having thus acquired an internal loop : as T_{11} diminishes, changes occur first as regards the lateral loops, and afterwards as regards the internal loop ; viz., each of the lateral loops is gradually pinched together until

$T_{11} = 1·80$? there are two new nodes Y′, Y″, each lateral loop being a figure of 8.

As T_{11} diminishes the figure-of-8-loop breaks up into a loop and oval, which oval continually diminishes until for

T = 1·10 ? the ovals have each become conjugate points, or there is a curve with two conjugate points *Z′*, *Z″*. As T_{11} diminishes the conjugate points have disappeared, and we have again a curve with an internal and two lateral loops ; but in the meantime the internal loop and the branch A′ A″ are continually approaching each other ; and, T = ·950, nodal isochronic, there is a node X on the axis. The curve consists of two figures of 8, each crossing itself at one of the points A′, A″, and the two crossing each other at the points A, X.

As T_{11} diminishes, the curve breaks off from X on each side of the axis so as not any longer to cross the axis (except at A), that is

T_{11} < ·950 ; curve is a chain intersecting itself at A′, A, A″ ; viz., from each loop there pass two branches, one inside, the other outside, the regulator, uniting themselves at A with the branches from the other loop outside and inside the regulator respectively ; and finally

$T_{11} = 0$, the curve is the arc A′AA″ of the regulator circle.

142. There is not in the several curves any discontinuity of direction at the point A′ or A″ : the branch from A within the shaded or half-shaded region, emerges at A′ or A″ into the lateral region, uniting itself with the upper branch of the loop ; it can only do this in virtue of its being at A′ or A″ a tangent to the separator (for otherwise it would cross the separator and regulator into the inner region) ; that is, the continuation thereof, or upper branch of the loop, must at the point A′ or A″ *touch the separator* ; it has been previously throughout assumed that this is so.

143. It is to be observed, both as regards the isoeccentric and the isochronic curves, that there is a real meaning in the obliterated portions ; viz., to any position of the orbit-pole on such obliterated portion of the curve there

corresponds a conic determined by means of a given trivector, but which, by reason of its being a convex hyperbola, or hyperbola such that the three points do not lie on the same branch thereof, is not regarded as an orbit. The obliterated portions have been in the present Memoir considered only so far as they present themselves in continuity with the curves which are the loci of the pole of a proper orbit, and for the purpose of explaining the course of these curves; and the curves completed as above are not the complete loci which would be obtained if, instead of the selected conic called the orbit, we had considered simultaneously the four conics determined by means of any given trivector; such extension of the theory would, it is probable, be interesting geometrically; but it would be devoid of all astronomical significance.

Guide Hyperbolas

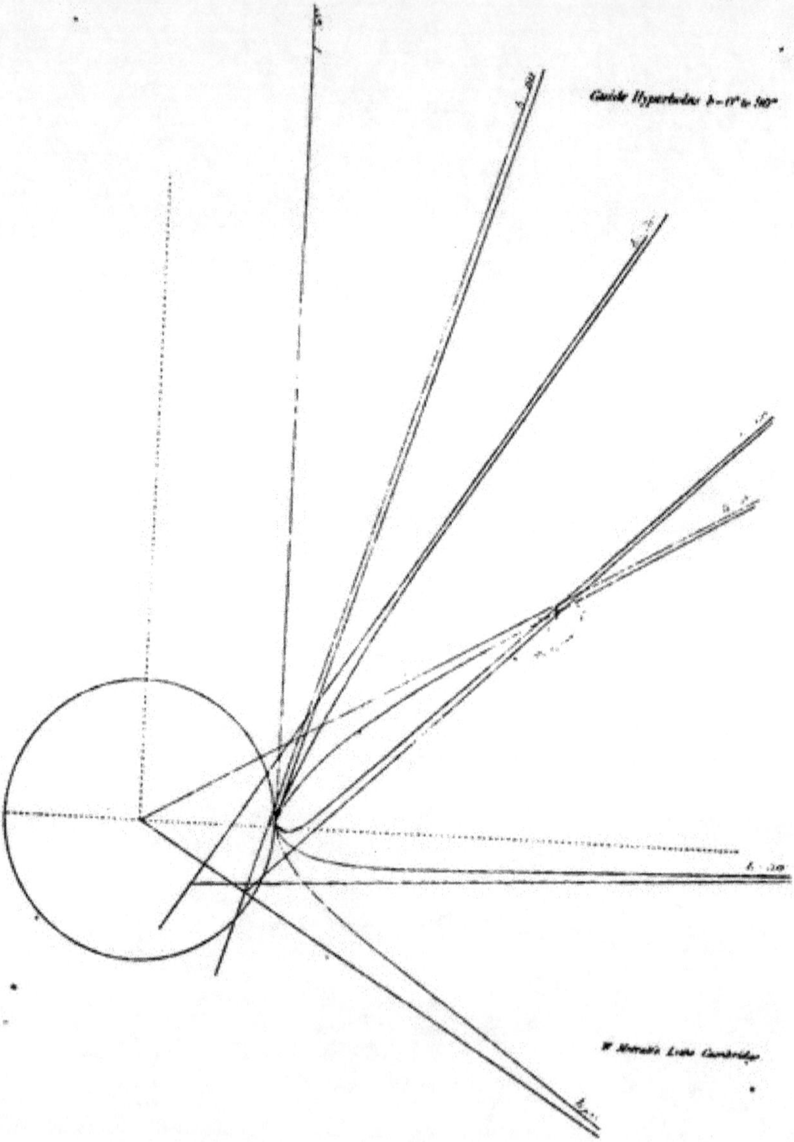

W. Metcalfe Litho Cambridge

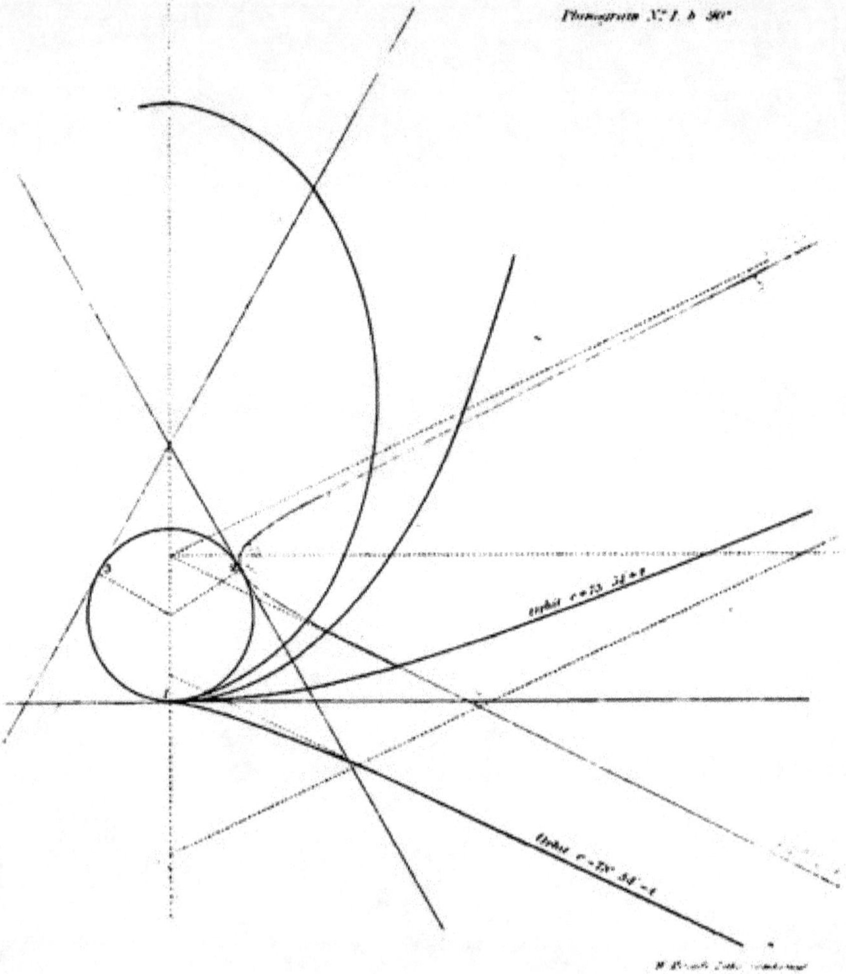

Planogram, No 2 & [illegible].

ε-Spherogram.

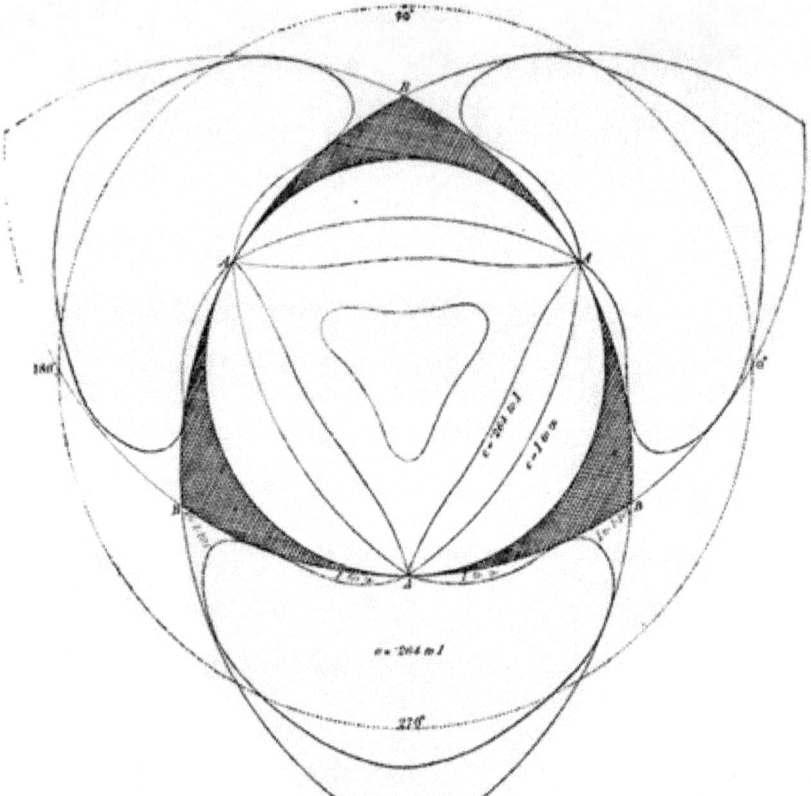

T_{12} — Sphærogram.

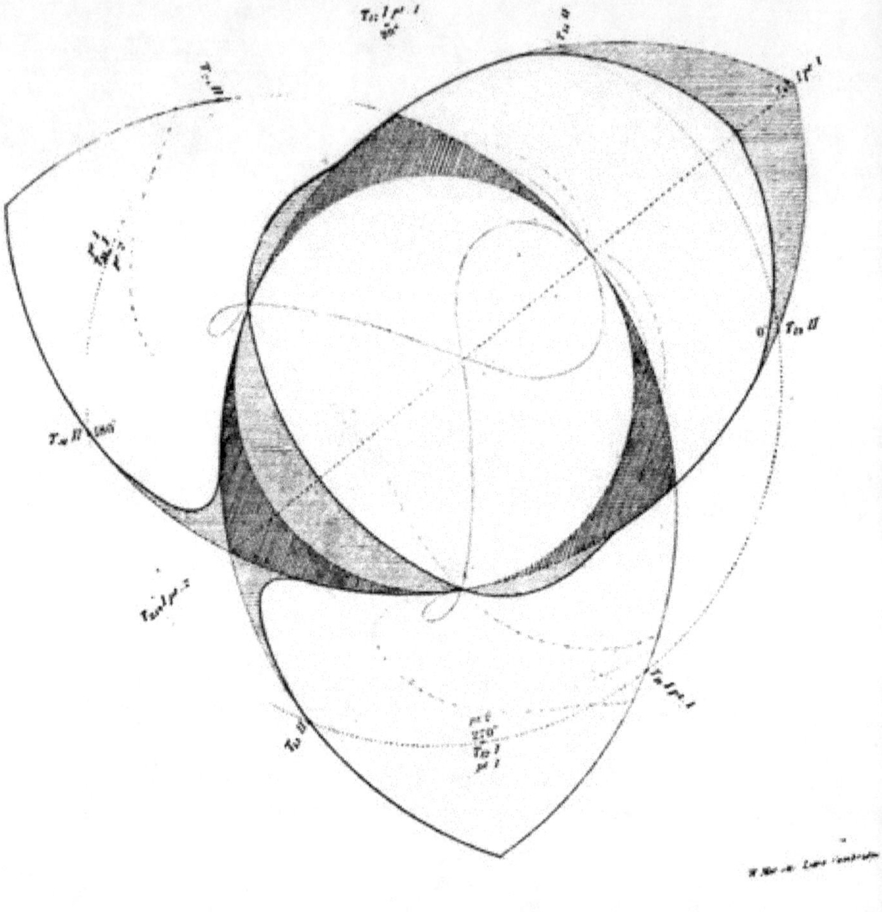

www.ingramcontent.com/pod-product-compliance
Lightning Source LLC
Chambersburg PA
CBHW021827190326
41518CB00007B/770
* 9 7 8 3 7 4 2 8 6 8 7 1 8 *